Math Sprints

Workbook 2

Tricia Salerno of SMARTTraining LLC

SM Singapore Math Inc®

Singapore Math Inc®

Copyright © 2010 Singapore Math Inc®

Published by
Singapore Math Inc
19535 SW 129th Ave.
Tualatin, OR 97062
U.S.A.
E-mail: customerservice@singaporemath.com
www.singaporemath.com

Written by
Tricia Salerno
SMARTTraining LLC
www.SingaporeMathTraining.com

Cover design by
Jopel Multimedia

Math Sprints Workbook 2
ISBN: 978-1-932906-37-0

First published 2010 in the United States of America
Reprinted 2013, 2014

Printed in Singapore

Introduction

"They don't know their facts!" This is the common lament we hear from teachers and parents around the United States. This book is here to help. Contained herein are math activities called "Sprints."

A sprint is a timed math test for FUN! That's right…There is no external pressure to achieve a certain score on sprints. The clear statement to your child that these are not for a grade should alleviate any math anxiety which sometimes arises during timed tests. It is important to let your child know that this is simply a competition against himself to improve mental math skills and it is for fun.

Act as if your child is actually involved in a race. Make it exciting. "On your mark, get set, GO!" Your child races to beat her own score and completes as many problems as possible in 60 seconds.

Importance of Math Facts

The importance of automatic recall of basic math facts has been argued in the past. In this day of technology, some say, why is it important to know the product of 6 and 8 when you can press a few buttons and have the answer quickly? In fact, you may have grown up with calculators in your hands and may have no idea how to help your child with mastering math facts because you don't know the facts yourself.

One of the problems with lack of automaticity with math facts is that if too much mental energy has to be spent recalling a basic fact, there's no mental energy left to solve the problem. Gersten and Chard stated:

> "Researchers explored the devastating effects of the lack of automaticity in several ways. Essentially they argued that the human mind has a limited capacity to process information, and if too much energy goes into figuring out what 9 plus 8 equals, little is left over to understand the concepts underlying multi-digit subtraction, long division, or complex multiplication." Gersten, R. and Chard, D. Number sense: Rethinking arithmetic instruction for students with mathematical disabilities. *Journal of Special Education* (1999), 3, 18–29 (1999).

Importance of Mental Math

Mental math is important for many reasons. Cathy Seeley, former president of the NCTM, stated:

> "Mental math is often associated with the ability to do computations quickly, but in its broadest sense, mental math also involves conceptual understanding and problem solving....Problem solving continues to be a high priority in school mathematics. Some argue that it is the most important mathematical goal for our students. Mental math provides both tools for solving problems and filters for evaluating answers. When a student has strong mental math skills, he or she can quickly test different approaches to a problem and determine whether the resulting path will lead toward a viable solution." (*NCTM News Bulletin*, December 2005).

Adrenaline

Research has proven that adrenaline aids memory. James McGaugh, a Professor of neurobiology at the University of California at Irvine, proved that adrenaline makes our brains remember better. When a sprint is given with a sense of urgency, as in a race, if your student experiences a rush of adrenaline, this can aid memory of the mental math being tested. It also makes the exercise significantly more fun!

About this Book

The Singapore Math curriculum stresses the use of mental math. These books are particularly useful to parents using Singapore Math material. In fact, sprints are useful to all parents interested in developing mental math fluency in their children.

These books were originally written for use in a classroom situation. They are reproduced here as a workbook for use in the home or in a setting with only a few students. The section below is an adaptation of how to give a sprint in a classroom situation. You can make further adaptations to meet your own child's needs, but be sure the keep it FUN! You will see that your child is racing to beat his own score each time he take one of these tests.

Each sprint is differentiated. The A sheet of each half of the sprint is easier than the B sheet. If you look closely at the A and B sheets of each sprint, the answers to the problems are the same. Many of the problems on the B sheets, however, require more mental calculation. If you have a child of average mental math abilities, you could give the A sheets first and the B sheets next, or later in the year, or not at all for that student. If you have a child that is strong in mental math abilities, you could give only the B sheet.

You may want to buy a sprint book at a level below the grade you teach so that your child gets used to taking sprints and feels very successful with them. Particularly if your child's mental math fluency is not where it should be, you can help her build it gradually by starting at a lower level.

If you are teaching more than one child, you may want to buy sprint books for each of their levels and administer the sprints to all of your children at the same time. You will have to read the answers separately for each child, though.

It is important to let your child know that this is simply a competition against himself to improve mental math skills and it is for fun. It is NOT for a grade.

How to Give a Sprint

1. Determine which sprint you want to give by looking at the topic of each. Each sprint has an A first half, an A second half, a B first half, and a B second half. The B sheets are for children stronger in math.

2. Give your child the workbook opened up to the page, face-down, of a "First Half" sheet for her to attempt to complete in 1 minute. Instruct your child not to turn the page face-up until told to "GO!" Get your child excited and enthuse: "On your mark, get set, GO!" and start your timer.

3. When the timer rings indicating one minute has elapsed, instruct your child to:

    a) stop working
    b) **draw a line under** the last problem completed
    c) put her pencil down.

4. Read the correct answers while your child pumps her hand in the air and respond "**yes**" to each problem that was answered correctly. Tell your child to mark the number of correct problems at the top of the page. If you are

administering the test to several children at different levels, you can have the other children complete the rest of the worksheet as you read the answers for one child.

5. Ask your child to complete the rest of the worksheet.

6. Let your child stand-up, stretch, run around, do jumping-jacks, etc.

7. Have your child sit back down and be ready to turn the page to do the reverse side, the second half.

8. Tell your child that the goal in this second half of the sprint is to beat his first score by at least one. **The child is competing only with himself.**

9. Repeat the preceding procedure through step 4, except that after making the correct number of problems at the top of the page, have your child compare the score on the first half to the score she got on the second half.

A good sprint:

1. Consists of two halves which test the same ONE skill. These are NOT random facts.
2. Builds in difficulty.
3. Is challenging enough that no one will be able to finish the first half in a minute.

NOTE: Look at each sprint and determine if your particular child can finish each half of a sprint in less than a minute. Some of the sprints have fewer problems than others. There is nothing wrong with doing a 30-second or 45-second sprint. Feel free to adjust the timing for your child, but be sure to keep the sense of urgency.

Acknowledgments

This series of books is due to the assistance of many people. Sprints are the brainchild of Dr. Yoram Sagher. Special thanks go to Ben Adler, Sam Adler, Laina Salerno and T. J. Salerno for their hours spent taking and re-taking the sprints contained herein. Linda West made it all come together. Thank you, thank you, thank you.

Math Sprints 2

1.	$10 + 8 =$	11.	$800 + 30 =$
2.	$50 + 6 =$	12.	$800 + 30 + 3 =$
3.	$100 + 20 + 3 =$	13.	$700 + 60 + 2 =$
4.	$200 + 50 + 5 =$	14.	$700 + 2 =$
5.	$400 + 80 =$	15.	$700 + 60 =$
6.	$100 + 70 + 3 =$	16.	$500 + 50 + 5 =$
7.	$100 + 60 =$	17.	$500 + 5 =$
8.	$300 + 70 + 9 =$	18.	$900 + 90 + 9 =$
9.	$500 + 60 + 4 =$	19.	$900 + 9 =$
10.	$800 + 3 =$	20.	$900 + 90 =$

201 A   Add.   Second Half

1.	10 + 6 =	11.	800 + 30 =
2.	40 + 6 =	12.	800 + 30 + 3 =
3.	100 + 10 + 4 =	13.	700 + 60 + 2 =
4.	200 + 60 + 8 =	14.	700 + 2 =
5.	300 + 20 =	15.	700 + 60 =
6.	300 + 70 + 4 =	16.	500 + 50 + 5 =
7.	200 + 50 =	17.	600 + 6 =
8.	300 + 70 + 9 =	18.	800 + 80 + 8 =
9.	500 + 60 + 4 =	19.	800 + 8 =
10.	800 + 3 =	20.	800 + 80 =

Math Sprints 2

| 201 B | Add. | First Half |

1. $4 + 6 + 8 =$	11. $500 + 300 + 10 + 20 =$
2. $25 + 25 + 6 =$	12. $250 + 300 + 250 + 33 =$
3. $80 + 20 + 20 + 3 =$	13. $200 + 250 + 250 + 62 =$
4. $50 + 100 + 100 + 5 =$	14. $250 + 450 + 2 =$
5. $160 + 240 + 80 =$	15. $380 + 380 =$
6. $90 + 70 + 10 + 3 =$	16. $420 + 110 + 20 + 5 =$
7. $20 + 40 + 60 + 40 =$	17. $50 + 45 + 105 + 300 + 5 =$
8. $270 + 30 + 70 + 9 =$	18. $333 + 333 + 333 =$
9. $250 + 250 + 60 + 4 =$	19. $150 + 250 + 500 + 9 =$
10. $300 + 300 + 200 + 3 =$	20. $300 + 310 + 320 + 60 =$

Math Sprints 2

1.	4 + 6 + 6 =	11.	500 + 300 + 10 + 20 =
2.	25 + 15 + 6 =	12.	250 + 300 + 250 + 33 =
3.	80 + 20 + 10 + 4 =	13.	200 + 250 + 250 + 62 =
4.	100 + 50 + 50 + 68 =	14.	250 + 450 + 2 =
5.	160 + 80 + 80 =	15.	380 + 380 =
6.	100 + 200 + 74 =	16.	420 + 110 + 20 +5 =
7.	50 + 100 + 100 =	17.	300 + 200 + 100 + 6 =
8.	270 + 30 + 70 + 9 =	18.	700 + 100 + 80 + 8 =
9.	250 + 250 + 60 + 4 =	19.	700 + 100 + 8 =
10.	300 + 300 + 200 + 3 =	20.	200 + 210 + 420 + 50 =

202 A		Write > or < on each line.	First Half
1.	9 _____ 10	16.	800 _____ 799
2.	90 _____ 100	17.	200 + 70 _____ 207
3.	200 _____ 199	18.	200 + 60 _____ 360
4.	201 _____ 210	19.	30 + 6 _____ 306
5.	253 _____ 235	20.	400 + 20 + 5 _____ 452
6.	301 _____ 309	21.	638 _____ 600 + 80
7.	499 _____ 500	22.	200 + 9 _____ 290
8.	980 _____ 908	23.	400 + 80 + 3 _____ 438
9.	765 _____ 756	24.	500 + 20 + 3 _____ 532
10.	354 _____ 543	25.	900 + 90 + 9 _____ 1000
11.	310 _____ 300	26.	5 + 10 + 100 _____ 151
12.	400 _____ 300	27.	8 + 20 _____ 208
13.	197 _____ 179	28.	315 _____ 300 + 5
14.	658 _____ 638	29.	691 _____ 1 + 80 + 600
15.	73 _____ 703	30.	1 + 20 + 700 ___ 700 + 20 + 2

202 A Write > or < on each line. Second Half

1.	8 _____ 10	16.	900 _____ 899
2.	80 _____ 100	17.	300 + 30 _____ 303
3.	200 _____ 199	18.	200 + 40 _____ 340
4.	302 _____ 320	19.	50 + 6 _____ 506
5.	343 _____ 335	20.	400 + 20 + 5 _____ 452
6.	401 _____ 409	21.	638 _____ 600 + 80
7.	499 _____ 510	22.	200 + 9 _____ 290
8.	770 _____ 707	23.	400 + 80 + 3 _____ 438
9.	765 _____ 756	24.	500 + 20 + 3 _____ 532
10.	354 _____ 543	25.	900 + 90 + 9 _____ 1000
11.	310 _____ 300	26.	5 + 10 + 100 _____ 151
12.	400 _____ 300	27.	8 + 20 _____ 208
13.	197 _____ 179	28.	315 _____ 300 + 5
14.	658 _____ 638	29.	491 _____ 1 + 80 + 400
15.	73 _____ 703	30.	1 + 30 + 500 _____ 500 + 30 + 2

| 202 B | Write > or < on each line. | First Half |

1.	9 _____ 10	16.	821 + 134 _____ 791 + 154
2.	90 _____ 100	17.	100 + 170 _____ 50 + 167
3.	90 + 9 _____ 90 + 8	18.	200 + 60 _____ 300 + 6 + 0
4.	300 − 99 _____ 190 + 20	19.	328 + 26 _____ 300 + 60
5.	180 + 80 _____ 230 − 24	20.	5 + 20 + 400 _____ 452
6.	300 − 96 _____ 140 + 120	21.	600 + 30 + 8 _____ 600 + 80 + 3
7.	700 + 20 + 3 _____ 700 + 30 + 2	22.	200 + 20 + 9 _____ 290
8.	100 + 18 + 8 _____ 100 + 11 + 8	23.	400 + 80 + 3 _____ 300 + 130 + 8
9.	300 − 40 _____ 190 + 16	24.	500 + 20 + 3 _____ 490 + 40 + 2
10.	90 + 5 _____ 905	25.	9 + 90 + 900 _____ 1000
11.	160 + 150 _____ 410 − 110	26.	5 + 10 + 100 _____ 50 + 50 + 50 + 1
12.	1000 − 110 _____ 900 − 20	27.	8 + 20 _____ 100 + 50 + 50 + 8
13.	100 + 90 + 7 _____ 100 + 70 + 9	28.	300 + 10 + 5 _____ 200 + 100 + 5
14.	1000 − 342 _____ 900 − 509	29.	600 + 90 + 1 _____ 1 + 30 + 50 + 600
15.	100 + 240 _____ 900 − 130	30.	1 + 20 + 700 _____ 700 + 20 + 2

202 B	Write > or < on each line.		**Second Half**

1.	8 _____ 10	16.	721 + 134 _____ 691 + 154
2.	80 _____ 100	17.	200 + 170 _____ 150 + 167
3.	80 + 8 _____ 80 + 7	18.	200 + 60 _____ 300 + 6 + 10
4.	400 − 99 _____ 290 + 20	19.	3210 + 26 _____ 3200 + 60
5.	160 + 60 _____ 230 − 24	20.	5 + 20 + 400 _____ 452
6.	300 − 96 _____ 140 + 120	21.	600 + 30 + 8 _____ 600 + 80 + 3
7.	700 + 20 + 3 _____ 700 + 30 + 2	22.	200 + 20 + 9 _____ 290
8.	100 + 18 + 8 _____ 100 + 11 + 8	23.	500 + 60 + 3 _____ 500 + 40 + 8
9.	300 − 40 _____ 190 + 16	24.	500 + 20 + 3 _____ 490 + 40 + 2
10.	90 + 5 _____ 905	25.	9 + 90 + 900 _____ 1000
11.	160 + 150 _____ 410 − 110	26.	5 + 10 + 100 _____ 50 + 50 + 50 + 1
12.	1000 − 110 _____ 900 − 20	27.	8 + 20 _____ 100 + 50 + 50 + 8
13.	100 + 90 + 7 _____ 100 + 70 + 9	28.	300 + 30 + 5 _____ 300 + 20 + 5
14.	1000 − 342 _____ 900 − 509	29.	600 + 90 + 1 _____ 1 + 30 + 50 + 600
15.	1000 − 240 _____ 900 − 130	30.	1 + 20 + 700 _____ 700 + 20 + 2

203 A   Fill in the blanks.   First Half

1.	1, 2, 3, _____	16.	700, 600, _____
2.	9, 8, _____	17.	800, 900, _____
3.	28, 29, _____	18.	105, 205, 305, _____
4.	72, 71, 70, _____	19.	110, 120, 130, _____
5.	123, 124, _____	20.	260, 250, _____
6.	298, 299, _____	21.	430, 330, 230, _____
7.	403, 402, 401, _____	22.	999, 899, 799, _____
8.	38, 39, _____	23.	587, 687, 787, _____
9.	64, 63, 62, 61, _____	24.	200, _____, 220, 230
10.	138, 139, _____	25.	480, 470, _____, 450
11.	262, 261, _____	26.	654, _____, 634, 624
12.	368, 369, _____	27.	701, 601, 501, _____
13.	490, 500, 510, _____	28.	810, _____, 610, 510
14.	620, 630, 640, _____	29.	297, 298, 299, _____
15.	100, 200, _____	30.	1000, _____, 980, 970

203 A Fill in the blanks. Second Half

1.	1, 2, 3, 4, _____	16.	405, 305, _____
2.	10, 9, _____	17.	800, 900, _____
3.	18, 19, _____	18.	102, 202, 302, _____
4.	42, 41, 40, _____	19.	110, 120, 130, _____
5.	114, 115, _____	20.	340, 330, 320, _____
6.	276, 275, 274, _____	21.	430, 330, 230, _____
7.	413, 412, 411, _____	22.	999, 899, 799, _____
8.	78, 79, _____	23.	587, 687, 787, _____
9.	84, 83, 82, 81, _____	24.	200, _____, 220, 230
10.	138, 139, _____	25.	480, 470, _____, 450
11.	262, 261, _____	26.	673, 663, 653, _____
12.	368, 369, _____	27.	701, 601, 501, _____
13.	490, 500, 510, _____	28.	810, _____, 610, 510
14.	620, 630, 640, _____	29.	297, 298, 299, _____
15.	100, 200, _____	30.	1000, _____, 980, 970

203 B	Fill in the blanks.	First Half

1.	10 , 8, 6, _____	16.	485, 490, 495, _____
2.	10 , 9, 8, _____	17.	400, 600, 800, _____
3.	10 , 20 , _____ , 40	18.	605, 505, _____ , 305
4.	65, 67, _____ , 71	19.	160, 150, _____ , 130
5.	115, 120, _____ , 130	20.	236, 238, _____
6.	200, 250, _____ , 350	21.	148, 142, 136, _____
7.	403, 402, 401, _____	22.	639, 669, _____ , 729
8.	36, 38, _____ , 42	23.	900, _____ , 874, 861
9.	64, 62, _____ , 58	24.	200, 205, _____ , 215
10.	120, 130, _____ , 150	25.	420, 440, _____ , 480
11.	220, 240, _____ , 280	26.	648, _____ , 640
12.	390, 380, _____ , 360	27.	397, 399, _____ , 403
13.	320, 420, _____ , 620	28.	705, _____ , 715
14.	665, 660, 655, _____	29.	288, _____ , 312, 324
15.	500, 400, _____	30.	996, 994, 992, _____

203 B   Fill in the blanks.   Second Half

1.	_____, 10, 15, 20	16.	185, 195, _____
2.	10 , 9, _____, 7	17.	400, 600, 800, _____
3.	10 , _____, 30, 40	18.	602, 502, _____, 302
4.	35, 37, _____, 41	19.	160, 150, _____, 130
5.	110, 112, 114, _____	20.	290, 300, _____, 320
6.	243, 253, 263, _____	21.	148, 142, 136, _____
7.	390, 400, _____, 420	22.	639, 669, _____, 729
8.	78, _____, 82	23.	900, _____, 874, 861
9.	_____, 78, 76	24.	200, 205, _____, 215
10.	120, 130, _____, 150	25.	420, 440, _____, 480
11.	220, 240, _____, 280	26.	646, _____, 640
12.	390, 380, _____, 360	27.	397, 399, _____, 403
13.	320, 420, _____, 620	28.	705, _____, 715
14.	665, 660, 655, _____	29.	288, _____, 312, 324
15.	500, 400, _____	30.	996, 994, 992, _____

204 A   Add.   First Half

1.	2 + 2 =	16.	51 + 31 =
2.	3 + 2 =	17.	52 + 32 =
3.	5 + 3 =	18.	52 + 42 =
4.	7 + 3 =	19.	62 + 30 =
5.	17 + 3 =	20.	62 + 33 =
6.	27 + 3 =	21.	70 + 30 =
7.	27 + 4 =	22.	72 + 30 =
8.	27 + 5 =	23.	70 + 32 =
9.	20 + 2 =	24.	100 + 100 =
10.	20 + 20 =	25.	200 + 200 =
11.	30 + 20 =	26.	300 + 201 =
12.	30 + 30 =	27.	300 + 225 =
13.	40 + 30 =	28.	303 + 225 =
14.	40 + 31 =	29.	313 + 225 =
15.	41 + 31 =	30.	376 + 124 =

Add. Second Half

1.	$2 + 1 =$	16.	$41 + 31 =$
2.	$2 + 2 =$	17.	$42 + 31 =$
3.	$2 + 3 =$	18.	$42 + 42 =$
4.	$6 + 3 =$	19.	$62 + 30 =$
5.	$16 + 3 =$	20.	$62 + 33 =$
6.	$26 + 3 =$	21.	$70 + 30 =$
7.	$26 + 4 =$	22.	$72 + 30 =$
8.	$26 + 5 =$	23.	$70 + 32 =$
9.	$20 + 2 =$	24.	$100 + 100 =$
10.	$20 + 20 =$	25.	$200 + 200 =$
11.	$30 + 20 =$	26.	$300 + 301 =$
12.	$30 + 30 =$	27.	$300 + 225 =$
13.	$40 + 20 =$	28.	$303 + 225 =$
14.	$40 + 21 =$	29.	$212 + 225 =$
15.	$41 + 21 =$	30.	$276 + 124 =$

Math Sprints 2

1.	$10 - 6 =$	16.	$200 - 118 =$
2.	$21 - 16 =$	17.	$201 - 117 =$
3.	$22 - 14 =$	18.	$203 - 109 =$
4.	$91 - 81 =$	19.	$200 - 108 =$
5.	$99 - 79 =$	20.	$204 - 109 =$
6.	$78 - 48 =$	21.	$210 - 110 =$
7.	$50 - 19 =$	22.	$301 - 199 =$
8.	$60 - 28 =$	23.	$401 - 299 =$
9.	$70 - 48 =$	24.	$962 - 762 =$
10.	$110 - 70 =$	25.	$887 - 487 =$
11.	$115 - 65 =$	26.	$702 - 201 =$
12.	$126 - 66 =$	27.	$1000 - 475 =$
13.	$135 - 65 =$	28.	$999 - 471 =$
14.	$121 - 50 =$	29.	$1000 - 462 =$
15.	$200 - 128 =$	30.	$987 - 487 =$

	204 B			Add.		Second Half

1.	$10 - 7 =$	16.	$200 - 128 =$
2.	$21 - 17 =$	17.	$200 - 127 =$
3.	$22 - 17 =$	18.	$208 - 124 =$
4.	$90 - 81 =$	19.	$200 - 108 =$
5.	$99 - 80 =$	20.	$204 - 109 =$
6.	$79 - 50 =$	21.	$210 - 110 =$
7.	$50 - 20 =$	22.	$301 - 199 =$
8.	$60 - 29 =$	23.	$401 - 299 =$
9.	$70 - 48 =$	24.	$962 - 762 =$
10.	$110 - 70 =$	25.	$887 - 487 =$
11.	$115 - 65 =$	26.	$702 - 101 =$
12.	$126 - 66 =$	27.	$1000 - 475 =$
13.	$135 - 75 =$	28.	$999 - 471 =$
14.	$122 - 61 =$	29.	$1000 - 563 =$
15.	$200 - 138 =$	30.	$987 - 587 =$

Subtract.

1.	$5 - 1 =$	16.	$56 - 23 =$
2.	$8 - 2 =$	17.	$56 - 33 =$
3.	$10 - 2 =$	18.	$56 - 43 =$
4.	$9 - 3 =$	19.	$56 - 54 =$
5.	$12 - 3 =$	20.	$60 - 50 =$
6.	$25 - 3 =$	21.	$68 - 53 =$
7.	$25 - 5 =$	22.	$78 - 53 =$
8.	$26 - 4 =$	23.	$78 - 56 =$
9.	$26 - 14 =$	24.	$88 - 67 =$
10.	$26 - 16 =$	25.	$98 - 68 =$
11.	$30 - 10 =$	26.	$98 - 70 =$
12.	$31 - 21 =$	27.	$25 - 25 =$
13.	$31 - 21 =$	28.	$50 - 25 =$
14.	$36 - 22 =$	29.	$75 - 25 =$
15.	$46 - 22 =$	30.	$100 - 25 =$

205 A Subtract. Second Half

1.	$4 - 1 =$	16.	$46 - 23 =$
2.	$8 - 2 =$	17.	$46 - 33 =$
3.	$9 - 2 =$	18.	$46 - 43 =$
4.	$9 - 3 =$	19.	$56 - 54 =$
5.	$11 - 3 =$	20.	$60 - 50 =$
6.	$25 - 3 =$	21.	$68 - 53 =$
7.	$15 - 5 =$	22.	$78 - 53 =$
8.	$16 - 4 =$	23.	$78 - 56 =$
9.	$16 - 14 =$	24.	$88 - 67 =$
10.	$16 - 16 =$	25.	$98 - 68 =$
11.	$40 - 10 =$	26.	$98 - 70 =$
12.	$31 - 21 =$	27.	$25 - 25 =$
13.	$31 - 21 =$	28.	$50 - 25 =$
14.	$36 - 22 =$	29.	$75 - 25 =$
15.	$46 - 22 =$	30.	$100 - 25 =$

205 B		Subtract.		First Half
1.	$6 - 2 =$	16.	$100 - 67 =$	
2.	$10 - 4 =$	17.	$90 - 67 =$	
3.	$12 - 4 =$	18.	$70 - 57 =$	
4.	$16 - 10 =$	19.	$201 - 199 =$	
5.	$20 - 11 =$	20.	$1000 - 990 =$	
6.	$33 - 11 =$	21.	$100 - 85 =$	
7.	$33 - 13 =$	22.	$310 - 285 =$	
8.	$40 - 18 =$	23.	$310 - 288 =$	
9.	$40 - 28 =$	24.	$310 - 289 =$	
10.	$100 - 90 =$	25.	$500 - 470 =$	
11.	$118 - 98 =$	26.	$608 - 580 =$	
12.	$109 - 99 =$	27.	$999 - 999 =$	
13.	$219 - 209 =$	28.	$100 - 75 =$	
14.	$206 - 192 =$	29.	$340 - 290 =$	
15.	$206 - 182 =$	30.	$415 - 340 =$	

205 B   Subtract.   Second Half

1.	$5 - 2 =$	16.	$100 - 77 =$
2.	$10 - 4 =$	17.	$90 - 77 =$
3.	$12 - 5 =$	18.	$70 - 67 =$
4.	$16 - 10 =$	19.	$201 - 199 =$
5.	$20 - 12 =$	20.	$1000 - 990 =$
6.	$33 - 11 =$	21.	$100 - 85 =$
7.	$33 - 23 =$	22.	$310 - 285 =$
8.	$30 - 18 =$	23.	$310 - 288 =$
9.	$30 - 28 =$	24.	$310 - 289 =$
10.	$90 - 90 =$	25.	$515 - 485 =$
11.	$118 - 88 =$	26.	$608 - 580 =$
12.	$109 - 99 =$	27.	$999 - 999 =$
13.	$219 - 209 =$	28.	$100 - 75 =$
14.	$206 - 192 =$	29.	$340 - 290 =$
15.	$206 - 182 =$	30.	$415 - 340 =$

206 A   Subtract.   First Half

1.	14 − 12 =	11.	730 − 220 =
2.	24 − 12 =	12.	797 − 687 =
3.	24 − 14 =	13.	142 − 131 =
4.	240 − 140 =	14.	748 − 436 =
5.	285 − 174 =	15.	999 − 899 =
6.	378 − 265 =	16.	888 − 700 =
7.	462 − 231 =	17.	709 − 609 =
8.	231 − 101 =	18.	738 − 518 =
9.	595 − 494 =	19.	684 − 503 =
10.	608 − 400 =	20.	856 − 632 =

206 A   Subtract.   Second Half

1.	$15 - 12 =$	11.	$640 - 220 =$
2.	$25 - 12 =$	12.	$597 - 487 =$
3.	$25 - 14 =$	13.	$152 - 132 =$
4.	$210 - 110 =$	14.	$748 - 526 =$
5.	$285 - 163 =$	15.	$999 - 899 =$
6.	$378 - 265 =$	16.	$888 - 700 =$
7.	$462 - 231 =$	17.	$709 - 609 =$
8.	$231 - 101 =$	18.	$738 - 518 =$
9.	$595 - 494 =$	19.	$684 - 503 =$
10.	$608 - 400 =$	20.	$856 - 632 =$

		Subtract.			First Half

206 B

1.	$114 - 112 =$		11.	$746 - 236 =$
2.	$224 - 212 =$		12.	$1000 - 890 =$
3.	$100 - 90 =$		13.	$200 - 189 =$
4.	$1000 - 900 =$		14.	$800 - 488 =$
5.	$285 - 174 =$		15.	$999 - 899 =$
6.	$378 - 265 =$		16.	$600 - 412 =$
7.	$462 - 231 =$		17.	$709 - 609 =$
8.	$341 - 211 =$		18.	$910 - 690 =$
9.	$595 - 494 =$		19.	$900 - 719 =$
10.	$618 - 410 =$		20.	$1000 - 776 =$

Subtract.

1.	$115 - 112 =$	11.	$646 - 226 =$
2.	$225 - 212 =$	12.	$1000 - 890 =$
3.	$101 - 90 =$	13.	$200 - 180 =$
4.	$800 - 700 =$	14.	$800 - 578 =$
5.	$365 - 243 =$	15.	$999 - 899 =$
6.	$278 - 165 =$	16.	$600 - 412 =$
7.	$462 - 231 =$	17.	$709 - 609 =$
8.	$341 - 211 =$	18.	$910 - 690 =$
9.	$595 - 494 =$	19.	$900 - 719 =$
10.	$618 - 410 =$	20.	$1000 - 776 =$

Add.

1.	$4 + 6 =$	16.	$150 + 150 =$
2.	$5 + 6 =$	17.	$250 + 250 =$
3.	$15 + 6 =$	18.	$255 + 255 =$
4.	$15 + 16 =$	19.	$300 + 309 =$
5.	$25 + 16 =$	20.	$301 + 309 =$
6.	$25 + 26 =$	21.	$311 + 309 =$
7.	$125 + 26 =$	22.	$500 + 500 =$
8.	$125 + 126 =$	23.	$490 + 510 =$
9.	$126 + 126 =$	24.	$480 + 520 =$
10.	$226 + 126 =$	25.	$480 + 519 =$
11.	$226 + 226 =$	26.	$489 + 511 =$
12.	$36 + 34 =$	27.	$578 + 23 =$
13.	$136 + 34 =$	28.	$578 + 123 =$
14.	$136 + 134 =$	29.	$578 + 133 =$
15.	$135 + 135 =$	30.	$578 + 199 =$

207 A		Add.		Second Half
1.	$4 + 7 =$	16.	$250 + 250 =$	
2.	$5 + 7 =$	17.	$350 + 250 =$	
3.	$15 + 7 =$	18.	$255 + 255 =$	
4.	$15 + 17 =$	19.	$300 + 305 =$	
5.	$25 + 17 =$	20.	$401 + 409 =$	
6.	$25 + 27=$	21.	$411 + 409 =$	
7.	$125 + 27 =$	22.	$500 + 500 =$	
8.	$125 + 127 =$	23.	$490 + 510 =$	
9.	$127 + 127 =$	24.	$480 + 520 =$	
10.	$227 + 127 =$	25.	$480 + 519 =$	
11.	$227 + 227 =$	26.	$489 + 511 =$	
12.	$36 + 34 =$	27.	$578 + 23 =$	
13.	$136 + 34 =$	28.	$578 + 123 =$	
14.	$136 + 134 =$	29.	$578 + 133 =$	
15.	$135 + 135 =$	30.	$578 + 199 =$	

© Singapore Math Inc®

207 B		Add.		First Half

1.	$4 + 6 =$	16.	$149 + 151 =$
2.	$5 + 6 =$	17.	$249 + 251 =$
3.	$15 + 6 =$	18.	$255 + 255 =$
4.	$13 + 18 =$	19.	$290 + 319 =$
5.	$25 + 16 =$	20.	$296 + 314 =$
6.	$25 + 26 =$	21.	$306 + 314 =$
7.	$26 + 125 =$	22.	$490 + 510 =$
8.	$125 + 126 =$	23.	$470 + 530 =$
9.	$116 + 136 =$	24.	$480 + 520 =$
10.	$216 + 136 =$	25.	$481 + 518 =$
11.	$216 + 236 =$	26.	$389 + 611 =$
12.	$26 + 44 =$	27.	$578 + 23 =$
13.	$136 + 34 =$	28.	$578 + 123 =$
14.	$136 + 134 =$	29.	$578 + 133 =$
15.	$115 + 155 =$	30.	$588 + 189 =$

1.	4 + 7 =	16.	249 + 251 =
2.	5 + 7 =	17.	249 + 351 =
3.	15 + 7 =	18.	255 + 255 =
4.	14 + 18 =	19.	290 + 315 =
5.	25 + 17 =	20.	296 + 514 =
6.	25 + 27 =	21.	306 + 514 =
7.	27 + 125 =	22.	490 + 510 =
8.	125 + 127 =	23.	470 + 530 =
9.	125 + 129 =	24.	480 + 520 =
10.	218 + 136 =	25.	481 + 518 =
11.	218 + 236 =	26.	389 + 611 =
12.	26 + 44 =	27.	578 + 23 =
13.	136 + 34 =	28.	578 + 123 =
14.	136 + 134 =	29.	578 + 133 =
15.	115 + 155 =	30.	588 + 189 =

207 B   Add.   Second Half

Math Sprints 2

1.	$10 - 6 =$	11.	$301 - 10 =$
2.	$20 - 6 =$	12.	$301 - 20 =$
3.	$20 - 7 =$	13.	$301 - 23 =$
4.	$120 - 7 =$	14.	$51 - 28 =$
5.	$18 - 9 =$	15.	$61 - 38 =$
6.	$118 - 9 =$	16.	$61 - 3 =$
7.	$118 - 19 =$	17.	$61 - 13 =$
8.	$200 - 5 =$	18.	$61 - 33 =$
9.	$200 - 15 =$	19.	$161 - 33 =$
10.	$200 - 35 =$	20.	$202 - 178 =$

Math Sprints 2

1.	$10 - 7 =$	11.	$303 - 10 =$
2.	$20 - 7 =$	12.	$303 - 20 =$
3.	$20 - 8 =$	13.	$303 - 23 =$
4.	$120 - 8 =$	14.	$61 - 28 =$
5.	$18 - 8 =$	15.	$71 - 38 =$
6.	$118 - 8 =$	16.	$71 - 3 =$
7.	$118 - 19 =$	17.	$61 - 13 =$
8.	$100 - 5 =$	18.	$61 - 33 =$
9.	$100 - 15 =$	19.	$161 - 33 =$
10.	$200 - 15 =$	20.	$202 - 178 =$

208 B		Subtract.	First Half

1.	$10 - 6 =$	11.	$301 - 10 =$
2.	$120 - 106 =$	12.	$301 - 20 =$
3.	$50 - 37 =$	13.	$576 - 298 =$
4.	$200 - 87 =$	14.	$900 - 877 =$
5.	$118 - 109 =$	15.	$901 - 878 =$
6.	$321 - 212 =$	16.	$61 - 3 =$
7.	$300 - 201 =$	17.	$61 - 13 =$
8.	$390 - 195 =$	18.	$61 - 33 =$
9.	$360 - 175 =$	19.	$261 - 133 =$
10.	$330 - 165 =$	20.	$202 - 178 =$

208 B	Subtract.	Second Half

1.	$10 - 7 =$	11.	$303 - 10 =$
2.	$120 - 107 =$	12.	$303 - 20 =$
3.	$50 - 38 =$	13.	$576 - 296 =$
4.	$200 - 88 =$	14.	$900 - 867 =$
5.	$118 - 108 =$	15.	$901 - 868 =$
6.	$321 - 211 =$	16.	$81 - 13 =$
7.	$300 - 201 =$	17.	$61 - 13 =$
8.	$400 - 305 =$	18.	$61 - 33 =$
9.	$400 - 315 =$	19.	$261 - 133 =$
10.	$330 - 145 =$	20.	$202 - 178 =$

Math Sprints 2

Fill in the blanks.

1.	1 m = _____ cm	11.	40 cm − 25 cm = _____ cm
2.	2 m = _____ cm	12.	50 cm − 25 cm = _____ cm
3.	3 m = _____ cm	13.	50 cm − 35 cm = _____ cm
4.	20 cm − 10 cm = _____ cm	14.	50 cm − 37 cm = _____ cm
5.	41 cm − 10 cm = _____ cm	15.	52 cm − 25 cm = _____ cm
6.	20 cm − 12 cm = _____ cm	16.	52 cm − 36 cm = _____ cm
7.	30 cm − 12 cm = _____ cm	17.	61 cm − 28 cm = _____ cm
8.	31 cm − 13 cm = _____ cm	18.	72 cm − 39 cm = _____ cm
9.	40 cm − 10 cm = _____ cm	19.	88 cm − 49 cm = _____ cm
10.	40 cm − 15 cm = _____ cm	20.	1 m − 25 cm = _____ cm

209 A   Fill in the blanks.   Second Half

1.	1 m = _____ cm	11.	40 cm – 35 cm = _____ cm
2.	2 m = _____ cm	12.	50 cm – 35 cm = _____ cm
3.	4 m = _____ cm	13.	50 cm – 25 cm = _____ cm
4.	30 cm – 10 cm = _____ cm	14.	50 cm – 37 cm = _____ cm
5.	31 cm – 10 cm = _____ cm	15.	52 cm – 25 cm = _____ cm
6.	20 cm – 12 cm = _____ cm	16.	52 cm – 36 cm = _____ cm
7.	40 cm – 12 cm = _____ cm	17.	61 cm – 28 cm = _____ cm
8.	41 cm – 13 cm = _____ cm	18.	72 cm – 49 cm = _____ cm
9.	40 cm – 10 cm = _____ cm	19.	88 cm – 39 cm = _____ cm
10.	40 cm – 15 cm = _____ cm	20.	1 m – 40 cm = _____ cm

209 B		Fill in the blanks.	**First Half**

1.	2m – 1 m = _____ cm	11.	70 cm – 55 cm = _____ cm
2.	3 m – 1m = _____ cm	12.	90 cm – 65 cm = _____ cm
3.	5 m – 2 m = _____ cm	13.	80 cm – 65 cm = _____ cm
4.	68 cm – 58 cm = _____ cm	14.	70 cm – 57 cm = _____ cm
5.	100 cm – 69 cm = _____ cm	15.	72 cm – 45 cm = _____ cm
6.	100 cm – 92 cm = _____ cm	16.	82 cm – 66 cm = _____ cm
7.	90 cm – 72 cm = _____ cm	17.	91 cm – 58 cm = _____ cm
8.	61 cm – 43 cm = _____ cm	18.	92 cm – 59 cm = _____ cm
9.	65 cm – 35 cm = _____ cm	19.	98 cm – 59 cm = _____ cm
10.	98 cm – 73 cm = _____ cm	20.	2 m – 1 m 25 cm = _____ cm

Math Sprints 2

1.	2m − 1 m = _____ cm	11.	70 cm − 65 cm = _____ cm
2.	3 m − 1m = _____ cm	12.	90 cm − 75 cm = _____ cm
3.	5 m − 1 m = _____ cm	13.	80 cm − 55 cm = _____ cm
4.	68 cm − 48 cm = _____ cm	14.	70 cm − 57 cm = _____ cm
5.	100 cm − 79 cm = _____ cm	15.	72 cm − 45 cm = _____ cm
6.	100 cm − 92 cm = _____ cm	16.	82 cm − 66 cm = _____ cm
7.	90 cm − 62 cm = _____ cm	17.	91 cm − 58 cm = _____ cm
8.	71 cm − 43 cm = _____ cm	18.	92 cm − 69 cm = _____ cm
9.	65 cm − 35 cm = _____ cm	19.	98 cm − 49 cm = _____ cm
10.	98 cm − 73 cm = _____ cm	20.	2 m − 1 m 40 cm = _____ cm

210 A   Fill in the blanks.   First Half

1.	1 yd = _____ ft	11.	12 ft = _____ yd
2.	2 yd = _____ ft	12.	9 ft = _____ yd
3.	3 yd = _____ ft	13.	15 ft = _____ yd
4.	4 yd = _____ ft	14.	24 in. = _____ ft
5.	5 yd = _____ ft	15.	48 in. = _____ ft
6.	1 ft = _____ in.	16.	36 in. = _____ ft
7.	2 ft = _____ in.	17.	1 yd = _____ in.
8.	3 ft = _____ in.	18.	2 yd = _____ in.
9.	4 ft = _____ in.	19.	72 in. = _____ yd
10.	6 ft = _____ in.	20.	144 in. = _____ yd

210 A   Fill in the blanks.   Second Half

1.	2 yd = _____ ft	11.	3 ft = _____ yd
2.	1 yd = _____ ft	12.	6 ft = _____ yd
3.	4 yd = _____ ft	13.	12 ft = _____ yd
4.	3 yd = _____ ft	14.	12 in. = _____ ft
5.	5 yd = _____ ft	15.	24 in. = _____ ft
6.	1 ft = _____ in.	16.	36 in. = _____ ft
7.	3 ft = _____ in.	17.	1 yd = _____ in.
8.	2 ft = _____ in.	18.	2 yd = _____ in.
9.	4 ft = _____ in.	19.	72 in. = _____ yd
10.	5 ft = _____ in.	20.	144 in. = _____ yd

Fill in the blanks.

1.	1 yd = _____ ft	11.	12 ft = _____ yd
2.	2 yd = _____ ft	12.	9 ft = _____ yd
3.	108 in. = _____ ft	13.	15 ft = _____ yd
4.	144 in. = _____ ft	14.	1 yd − 12 in. = _____ ft
5.	5 yd = _____ ft	15.	1 ft − 8 in. = _____ in.
6.	1 yd − 24 in. = _____ in.	16.	36 in. = _____ ft
7.	1 yd − 1 ft = _____ in.	17.	6 ft − 1 yd = _____ in.
8.	3 ft = _____ in.	18.	2 yd = _____ in.
9.	1 yd + 1 ft = _____ in.	19.	72 in. = _____ yd
10.	1 yd + 1 ft + 2 ft = _____ in.	20.	144 in. = _____ yd

1.	2 yd = _____ ft	11.	12 in. = _____ ft
2.	1 yd = _____ ft	12.	6 ft = _____ yd
3.	144 in. = _____ ft	13.	12 ft = _____ yd
4.	108 in. = _____ ft	14.	1 yd – 24 in. = _____ ft
5.	5 yd = _____ ft	15.	1 ft – 10 in. = _____ in.
6.	1 yd – 24 in. = _____ in.	16.	36 in. = _____ ft
7.	1 yd = _____ in.	17.	6 ft – 1 yd = _____ in.
8.	2 ft = _____ in.	18.	2 yd = _____ in.
9.	1 yd + 1 ft = _____ in.	19.	72 in. = _____ yd
10.	1 yd + 1 ft + 1 ft = _____ in.	20.	144 in. = _____ yd

211 A		Multiply.	First Half
1.	$2 \times 1 =$	16.	$10 \times 2 =$
2.	$1 \times 2 =$	17.	$2 \times 10 =$
3.	$2 \times 2 =$	18.	$2 \times 4 =$
4.	$2 \times 3 =$	19.	$5 \times 2 =$
5.	$3 \times 2 =$	20.	$10 \times 2 =$
6.	$4 \times 2 =$	21.	$9 \times 2 =$
7.	$2 \times 5 =$	22.	$2 \times 8 =$
8.	$2 \times 6 =$	23.	$2 \times 7 =$
9.	$6 \times 2 =$	24.	$2 \times 6 =$
10.	$2 \times 7 =$	25.	$2 \times 4 =$
11.	$7 \times 2 =$	26.	$3 \times 2 =$
12.	$8 \times 2 =$	27.	$2 \times 2 =$
13.	$2 \times 8 =$	28.	$2 \times 0 =$
14.	$2 \times 9 =$	29.	$0 \times 2 =$
15.	$9 \times 2 =$	30.	$2 \times 8 =$

211 A   Multiply.   Second Half

1.	$2 \times 0 =$	16.	$10 \times 2 =$
2.	$0 \times 2 =$	17.	$2 \times 10 =$
3.	$2 \times 1 =$	18.	$2 \times 4 =$
4.	$2 \times 2 =$	19.	$5 \times 2 =$
5.	$2 \times 3 =$	20.	$10 \times 2 =$
6.	$4 \times 2 =$	21.	$9 \times 2 =$
7.	$2 \times 4 =$	22.	$2 \times 8 =$
8.	$6 \times 2 =$	23.	$2 \times 7 =$
9.	$2 \times 6 =$	24.	$2 \times 6 =$
10.	$2 \times 7 =$	25.	$2 \times 4 =$
11.	$7 \times 2 =$	26.	$3 \times 2 =$
12.	$8 \times 2 =$	27.	$2 \times 2 =$
13.	$2 \times 8 =$	28.	$2 \times 0 =$
14.	$2 \times 9 =$	29.	$0 \times 2 =$
15.	$9 \times 2 =$	30.	$2 \times 8 =$

211 B   Fill in the blanks.   First Half

1.	How many fours are in 8? _____	16.	How many twos are in 40? _____
2.	How many fives are in 10? _____	17.	$2 \times 10 =$ _____
3.	How many fives are in 20? _____	18.	How many twos are in 16? _____
4.	How many fives are in 30? _____	19.	$5 \times 2 =$ _____
5.	How many twos are in 12? _____	20.	$10 \times 2 =$ _____
6.	How many tens are in 80? _____	21.	$9 \times 2 =$ _____
7.	How many twos are in 20? _____	22.	$2 \times 8 =$ _____
8.	How many twos are in 24? _____	23.	$2 \times 7 =$ _____
9.	$6 \times 2 =$ _____	24.	$2 \times 6 =$ _____
10.	$2 \times 7 =$ _____	25.	$2 \times 2 \times 2 =$ _____
11.	$7 \times 2 =$ _____	26.	How many twos are in 12? _____
12.	$8 \times 2 =$ _____	27.	$2 \times 2 =$ _____
13.	$2 \times 8 =$ _____	28.	$2 \times 0 =$ _____
14.	$2 \times 9 =$ _____	29.	$0 \times 2 =$ _____
15.	$9 \times 2 =$ _____	30.	$2 \times 2 \times 2 \times 2 =$ _____

211 B   Fill in the blanks.   Second Half

1.	$24 \times 0 =$ _____	16.	How many twos are in 40? _____
2.	$0 \times 76 =$ _____	17.	$2 \times 10 =$ _____
3.	How many fives are in 10? _____	18.	How many twos are in 16? _____
4.	How many fives are in 20? _____	19.	$5 \times 2 =$ _____
5.	How many fives are in 30? _____	20.	$10 \times 2 =$ _____
6.	How many twos are in 16? _____	21.	$9 \times 2 =$ _____
7.	How many fives are in 40? _____	22.	$2 \times 8 =$ _____
8.	How many fives are in 60? _____	23.	$2 \times 7 =$ _____
9.	$6 \times 2 =$ _____	24.	$2 \times 6 =$ _____
10.	$2 \times 7 =$ _____	25.	$2 \times 2 \times 2 =$ _____
11.	$7 \times 2 =$ _____	26.	How many twos are in 12? _____
12.	$8 \times 2 =$ _____	27.	$2 \times 2 =$ _____
13.	$2 \times 8 =$ _____	28.	$2 \times 0 =$ _____
14.	$2 \times 9 =$ _____	29.	$0 \times 2 =$ _____
15.	$9 \times 2 =$ _____	30.	$2 \times 2 \times 2 \times 2 =$ _____

212 A   Multiply or divide.   First Half

1.	$3 \times 1 =$	16.	$3 \times 0 =$
2.	$3 \times 2 =$	17.	$3 \times 10 =$
3.	$1 \times 3 =$	18.	$3 \times 9 =$
4.	$2 \times 3 =$	19.	$3 \times 7 =$
5.	$3 \times 3 =$	20.	$3 \times 5 =$
6.	$4 \times 3 =$	21.	$3 \times 4 =$
7.	$6 \times 3 =$	22.	$3 \times 6 =$
8.	$5 \times 3 =$	23.	$30 \div 3 =$
9.	$3 \times 5 =$	24.	$27 \div 3 =$
10.	$3 \times 6 =$	25.	$21 \div 3 =$
11.	$7 \times 3 =$	26.	$15 \div 3 =$
12.	$8 \times 3 =$	27.	$9 \div 3 =$
13.	$3 \times 7 =$	28.	$24 \div 3 =$
14.	$3 \times 9 =$	29.	$18 \div 3 =$
15.	$10 \times 3 =$	30.	$3 \div 3 =$

212 A Multiply or divide. Second Half

1.	3 x 0 =	16.	3 x 0 =
2.	3 x 1 =	17.	3 x 10 =
3.	0 x 3 =	18.	3 x 9 =
4.	1 x 3 =	19.	3 x 7 =
5.	3 x 2 =	20.	3 x 5 =
6.	3 x 4 =	21.	3 x 4 =
7.	7 x 3 =	22.	3 x 6 =
8.	6 x 3 =	23.	30 ÷ 3 =
9.	3 x 6 =	24.	27 ÷ 3 =
10.	3 x 5 =	25.	21 ÷ 3 =
11.	6 x 3 =	26.	15 ÷ 3 =
12.	7 x 3 =	27.	9 ÷ 3 =
13.	3 x 6 =	28.	24 ÷ 3 =
14.	3 x 8 =	29.	18 ÷ 3 =
15.	9 x 3 =	30.	3 ÷ 3 =

212 B Multiply or divide. First Half

1.	$3 \times 1 =$	16.	$3 \times 0 =$
2.	$3 \times 2 =$	17.	$90 \div 3 =$
3.	$9 \div 3 =$	18.	$3 \times 9 =$
4.	$18 \div 3 =$	19.	$3 \times 7 =$
5.	$27 \div 3 =$	20.	$3 \times 5 =$
6.	$4 \times 3 =$	21.	$3 \times 4 =$
7.	$6 \times 3 =$	22.	$3 \times 6 =$
8.	$5 \times 3 =$	23.	$30 \div 3 =$
9.	$3 \times 5 =$	24.	$27 \div 3 =$
10.	$3 \times 6 =$	25.	$21 \div 3 =$
11.	$7 \times 3 =$	26.	$15 \div 3 =$
12.	$8 \times 3 =$	27.	$9 \div 3 =$
13.	$3 \times 7 =$	28.	$24 \div 3 =$
14.	$3 \times 3 \times 3 =$	29.	$18 \div 3 =$
15.	$10 \times 3 =$	30.	$3 \div 3 =$

212 B   Multiply or divide.   Second Half

1.	$3 \times 0 =$	16.	$3 \times 0 =$
2.	$3 \times 1 =$	17.	$90 \div 3 =$
3.	$0 \times 3 =$	18.	$3 \times 9 =$
4.	$9 \div 3 =$	19.	$3 \times 7 =$
5.	$18 \div 3 =$	20.	$3 \times 5 =$
6.	$4 \times 3 =$	21.	$3 \times 4 =$
7.	$7 \times 3 =$	22.	$3 \times 6 =$
8.	$3 \times 6 =$	23.	$30 \div 3 =$
9.	$3 \times 3 \times 2 =$	24.	$27 \div 3 =$
10.	$3 \times 5 =$	25.	$21 \div 3 =$
11.	$6 \times 3 =$	26.	$15 \div 3 =$
12.	$7 \times 3 =$	27.	$9 \div 3 =$
13.	$3 \times 6 =$	28.	$24 \div 3 =$
14.	$3 \times 2 \times 2 \times 2 =$	29.	$18 \div 3 =$
15.	$9 \times 3 =$	30.	$3 \div 3 =$

213 A   Multiply.   First Half

1.	$1 \times 2 =$	16.	$3 \times 8 =$
2.	$2 \times 2 =$	17.	$2 \times 7 =$
3.	$3 \times 3 =$	18.	$2 \times 6 =$
4.	$3 \times 2 =$	19.	$3 \times 6 =$
5.	$3 \times 1 =$	20.	$3 \times 7 =$
6.	$3 \times 4 =$	21.	$3 \times 9 =$
7.	$4 \times 3 =$	22.	$9 \times 3 =$
8.	$3 \times 6 =$	23.	$9 \times 2 =$
9.	$5 \times 3 =$	24.	$10 \times 2 =$
10.	$5 \times 2 =$	25.	$10 \times 3 =$
11.	$6 \times 2 =$	26.	$9 \times 3 =$
12.	$4 \times 2 =$	27.	$7 \times 3 =$
13.	$7 \times 2 =$	28.	$3 \times 8 =$
14.	$7 \times 3 =$	29.	$3 \times 6 =$
15.	$2 \times 8 =$	30.	$6 \times 2 =$

213 A Multiply. Second Half

1.	$1 \times 1 =$	16.	$3 \times 6 =$
2.	$2 \times 1 =$	17.	$2 \times 6 =$
3.	$2 \times 2 =$	18.	$2 \times 7 =$
4.	$3 \times 2 =$	19.	$3 \times 7 =$
5.	$3 \times 4 =$	20.	$3 \times 6 =$
6.	$3 \times 3 =$	21.	$3 \times 8 =$
7.	$5 \times 3 =$	22.	$9 \times 3 =$
8.	$3 \times 7 =$	23.	$9 \times 2 =$
9.	$6 \times 3 =$	24.	$10 \times 2 =$
10.	$6 \times 2 =$	25.	$10 \times 3 =$
11.	$7 \times 2 =$	26.	$9 \times 3 =$
12.	$9 \times 2 =$	27.	$7 \times 3 =$
13.	$9 \times 3 =$	28.	$3 \times 8 =$
14.	$8 \times 3 =$	29.	$3 \times 6 =$
15.	$2 \times 8 =$	30.	$6 \times 2 =$

213 B Multiply or divide. First Half

1.	$4 \div 2 =$	16.	$3 \times 8 =$
2.	$8 \div 2 =$	17.	$2 \times 7 =$
3.	$27 \div 3 =$	18.	$2 \times 3 \times 2 =$
4.	$18 \div 3 =$	19.	$3 \times 6 =$
5.	$9 \div 3 =$	20.	$7 \times 3 =$
6.	$36 \div 3 =$	21.	$3 \times 3 \times 3 =$
7.	$2 \times 2 \times 3 =$	22.	$9 \times 3 =$
8.	$3 \times 6 =$	23.	$3 \times 6 =$
9.	$5 \times 3 =$	24.	$2 \times 2 \times 5 =$
10.	$30 \div 3 =$	25.	$10 \times 3 =$
11.	$3 \times 2 \times 2 =$	26.	$9 \times 3 =$
12.	$24 \div 3 =$	27.	$3 \times 7 =$
13.	$7 \times 2 =$	28.	$3 \times 8 =$
14.	$7 \times 3 =$	29.	$3 \times 6 =$
15.	$2 \times 2 \times 2 \times 2 =$	30.	$3 \times 2 \times 2 =$

213 B   Multiply or divide.   Second Half

1.	$4 \div 4 =$	16.	$3 \times 6 =$
2.	$8 \div 4 =$	17.	$2 \times 2 \times 3 =$
3.	$12 \div 3 =$	18.	$28 \div 2 =$
4.	$18 \div 3 =$	19.	$42 \div 2 =$
5.	$6 \times 2 =$	20.	$6 \times 3 =$
6.	$36 \div 4 =$	21.	$3 \times 2 \times 2 \times 2 =$
7.	$5 \times 3 =$	22.	$9 \times 3 =$
8.	$3 \times 7 =$	23.	$3 \times 6 =$
9.	$6 \times 3 =$	24.	$2 \times 2 \times 5 =$
10.	$36 \div 3 =$	25.	$10 \times 3 =$
11.	$7 \times 2 =$	26.	$9 \times 3 =$
12.	$3 \times 3 \times 2 =$	27.	$3 \times 7 =$
13.	$9 \times 3 =$	28.	$3 \times 8 =$
14.	$8 \times 3 =$	29.	$3 \times 6 =$
15.	$2 \times 2 \times 2 \times 2 =$	30.	$3 \times 2 \times 2 =$

214 A   Fill in the blanks.   First Half

1.	$3 + \underline{\hspace{2cm}} = 4$	11.	$20 - \underline{\hspace{2cm}} = 5$
2.	$4 - \underline{\hspace{2cm}} = 1$	12.	$21 - \underline{\hspace{2cm}} = 6$
3.	$5 + \underline{\hspace{2cm}} = 7$	13.	$\underline{\hspace{2cm}} - 5 = 10$
4.	$7 - 5 = \underline{\hspace{2cm}}$	14.	$20 - \underline{\hspace{2cm}} = 13$
5.	$6 + \underline{\hspace{2cm}} = 10$	15.	$30 - \underline{\hspace{2cm}} = 13$
6.	$10 - \underline{\hspace{2cm}} = 6$	16.	$13 + \underline{\hspace{2cm}} = 30$
7.	$12 - \underline{\hspace{2cm}} = 10$	17.	$14 + \underline{\hspace{2cm}} = 30$
8.	$12 - 10 = \underline{\hspace{2cm}}$	18.	$\underline{\hspace{2cm}} - 8 = 10$
9.	$8 + \underline{\hspace{2cm}} = 14$	19.	$\underline{\hspace{2cm}} - 18 = 40$
10.	$14 - \underline{\hspace{2cm}} = 8$	20.	$\underline{\hspace{2cm}} + 14 = 30$

214 A Fill in the blanks. Second Half

1.	$2 + \underline{\hspace{2cm}} = 3$	11.	$20 - \underline{\hspace{2cm}} = 5$
2.	$3 - \underline{\hspace{2cm}} = 1$	12.	$21 - \underline{\hspace{2cm}} = 6$
3.	$4 + \underline{\hspace{2cm}} = 6$	13.	$\underline{\hspace{2cm}} - 5 = 10$
4.	$6 - 4 = \underline{\hspace{2cm}}$	14.	$20 - \underline{\hspace{2cm}} = 13$
5.	$5 + \underline{\hspace{2cm}} = 10$	15.	$30 - \underline{\hspace{2cm}} = 13$
6.	$10 - \underline{\hspace{2cm}} = 5$	16.	$13 + \underline{\hspace{2cm}} = 30$
7.	$13 - \underline{\hspace{2cm}} = 10$	17.	$14 + \underline{\hspace{2cm}} = 30$
8.	$13 - 10 = \underline{\hspace{2cm}}$	18.	$\underline{\hspace{2cm}} - 8 = 10$
9.	$9 + \underline{\hspace{2cm}} = 14$	19.	$\underline{\hspace{2cm}} - 18 = 40$
10.	$14 - \underline{\hspace{2cm}} = 9$	20.	$\underline{\hspace{2cm}} + 14 = 30$

214 B   Fill in the blanks.   First Half

1.	$13 + \underline{\hspace{2cm}} = 14$	11.	$20 - \underline{\hspace{2cm}} = 5$
2.	$14 - \underline{\hspace{2cm}} = 11$	12.	$21 - \underline{\hspace{2cm}} = 6$
3.	$15 + \underline{\hspace{2cm}} = 17$	13.	$205 + \underline{\hspace{2cm}} = 220$
4.	$27 - 25 = \underline{\hspace{2cm}}$	14.	$198 + \underline{\hspace{2cm}} = 205$
5.	$6 + \underline{\hspace{2cm}} = 10$	15.	$189 - \underline{\hspace{2cm}} = 172$
6.	$62 - \underline{\hspace{2cm}} = 58$	16.	$201 - \underline{\hspace{2cm}} = 184$
7.	$13 + \underline{\hspace{2cm}} = 15$	17.	$\underline{\hspace{2cm}} + 495 = 511$
8.	$21 - \underline{\hspace{2cm}} = 19$	18.	$\underline{\hspace{2cm}} + 389 = 407$
9.	$45 - \underline{\hspace{2cm}} = 39$	19.	$402 - \underline{\hspace{2cm}} = 344$
10.	$28 + \underline{\hspace{2cm}} = 34$	20.	$\underline{\hspace{2cm}} + 96 = 112$

214 B   Fill in the blanks.   Second Half

1.	13 + _____ = 14	11.	20 − _____ = 5
2.	13 − _____ = 11	12.	21 − _____ = 6
3.	14 + _____ = 16	13.	205 + _____ = 220
4.	37 − 35 = _____	14.	198 + _____ = 205
5.	6 + _____ = 11	15.	189 − _____ = 172
6.	62 − _____ = 57	16.	201 − _____ = 184
7.	13 + _____ = 16	17.	_____ + 495 = 511
8.	21 − _____ = 18	18.	_____ + 389 = 407
9.	44 − _____ = 39	19.	402 − _____ = 344
10.	30 + _____ = 35	20.	_____ + 96 = 112

215 A   Add.   First Half

1.	127 + 2 =	11.	60 + 50 =
2.	127 + 20 =	12.	40 + 50 =
3.	127 + 200 =	13.	42 + 50 =
4.	24 + 3 =	14.	42 + 40 =
5.	24 + 30 =	15.	80 + 40 =
6.	10 + 30 =	16.	90 + 30 =
7.	30 + 20 =	17.	90 + 50 =
8.	50 + 40 =	18.	82 + 40 =
9.	60 + 20 =	19.	86 + 60 =
10.	60 + 40 =	20.	96 + 60 =

215 A   Add.   Second Half

1.	$125 + 2 =$	11.	$60 + 50 =$
2.	$125 + 20 =$	12.	$40 + 50 =$
3.	$125 + 200 =$	13.	$42 + 50 =$
4.	$26 + 3 =$	14.	$42 + 40 =$
5.	$26 + 30 =$	15.	$80 + 40 =$
6.	$20 + 50 =$	16.	$90 + 30 =$
7.	$50 + 20 =$	17.	$90 + 50 =$
8.	$50 + 30 =$	18.	$82 + 40 =$
9.	$60 + 30 =$	19.	$88 + 60 =$
10.	$60 + 40 =$	20.	$98 + 60 =$

215 B   Add.   First Half

1.	127 + 2 =	11.	80 + 30 =
2.	127 + 20 =	12.	35 + 25 + 30 =
3.	127 + 200 =	13.	30 + 30 + 32 =
4.	24 + 3 =	14.	20 + 30 + 32 =
5.	24 + 30 =	15.	60 + 35 + 25 =
6.	13 + 17 + 10 =	16.	80 + 15 + 25 =
7.	18 + 20 + 12 =	17.	90 + 50 =
8.	30 + 30 + 30 =	18.	82 + 40 =
9.	20 + 20 + 20 + 20 =	19.	86 + 60 =
10.	57 + 43 =	20.	80 + 50 + 16 + 10 =

215 B   Add.   Second Half

1.	125 + 2 =	11.	80 + 30 =
2.	125 + 20 =	12.	35 + 25 + 30 =
3.	125 + 200 =	13.	30 + 30 + 32 =
4.	24 + 5 =	14.	20 + 30 + 32 =
5.	26 + 30 =	15.	60 + 35 + 25 =
6.	13 + 17 + 40 =	16.	80 + 15 + 25 =
7.	18 + 40 + 12 =	17.	90 + 50 =
8.	30 + 20 + 30 =	18.	82 + 40 =
9.	20 + 30 + 10 + 30 =	19.	88 + 60 =
10.	47 + 53 =	20.	80 + 50 + 18 + 10 =

216 A   Add.   First Half

1.	$9 + 9 =$	11.	$168 + 9 =$
2.	$14 + 9 =$	12.	$168 + 19 =$
3.	$23 + 9 =$	13.	$168 + 28 =$
4.	$136 + 9 =$	14.	$605 + 8 =$
5.	$139 + 5 =$	15.	$224 + 50 =$
6.	$139 + 15 =$	16.	$321 + 300 =$
7.	$138 + 7 =$	17.	$404 + 400 =$
8.	$138 + 27 =$	18.	$404 + 9 =$
9.	$167 + 4 =$	19.	$408 + 9 =$
10.	$167 + 24 =$	20.	$418 + 9 =$

216 A Add. Second Half

1.	8 + 9 =	11.	157 + 9 =
2.	13 + 9 =	12.	157 + 19 =
3.	33 + 9 =	13.	168 + 28 =
4.	133 + 9 =	14.	605 + 8 =
5.	139 + 6 =	15.	224 + 50 =
6.	139 + 16 =	16.	321 + 300 =
7.	139 + 7 =	17.	404 + 400 =
8.	139 + 27 =	18.	404 + 9 =
9.	148 + 4 =	19.	408 + 9 =
10.	148 + 14 =	20.	418 + 9 =

216 B   Add.   First Half

1.	9 + 9 =	11.	168 + 9 =
2.	14 + 9 =	12.	158 + 29 =
3.	23 + 9 =	13.	168 + 28 =
4.	129 + 16 =	14.	598 + 15 =
5.	118 + 26 =	15.	224 + 50 =
6.	139 + 15 =	16.	200 + 200 + 221 =
7.	138 + 7 =	17.	300 + 300 + 204 =
8.	138 + 27 =	18.	9 + 404 =
9.	167 + 4 =	19.	9 + 408 =
10.	157 + 34 =	20.	9 + 418 =

216 B	Add.	Second Half

1.	8 + 9 =	11.	157 + 9 =
2.	14 + 8 =	12.	147 + 29 =
3.	33 + 9 =	13.	168 + 28 =
4.	133 + 9 =	14.	598 + 15 =
5.	118 + 27 =	15.	224 + 50 =
6.	139 + 16 =	16.	200 + 200 + 221 =
7.	138 + 8 =	17.	300 + 300 + 204 =
8.	138 + 28 =	18.	9 + 404 =
9.	148 + 4 =	19.	9 + 408 =
10.	147 + 15 =	20.	9 + 418 =

217 A		Add.		First Half
1.	13 + 3 =	16.	83 + 15 =	
2.	14 + 5 =	17.	26 + 7 =	
3.	24 + 5 =	18.	26 + 17 =	
4.	61 + 22 =	19.	36 + 17 =	
5.	42 + 17 =	20.	36 + 18 =	
6.	71 + 18 =	21.	36 + 28 =	
7.	53 + 22 =	22.	37 + 30 =	
8.	72 + 15 =	23.	37 + 29 =	
9.	72 + 17 =	24.	38 + 22 =	
10.	82 + 17 =	25.	38 + 25 =	
11.	90 + 8 =	26.	26 + 38 =	
12.	88 + 10 =	27.	27 + 39 =	
13.	86 + 12 =	28.	31 + 42 =	
14.	84 + 14 =	29.	39 + 42 =	
15.	82 + 16 =	30.	39 + 55 =	

217 A   Add.   Second Half

1.	12 + 3 =	16.	72 + 14 =
2.	13 + 5 =	17.	24 + 7 =
3.	23 + 5 =	18.	24 + 17 =
4.	62 + 23 =	19.	34 + 17 =
5.	41 + 18 =	20.	34 + 18 =
6.	72 + 18 =	21.	34 + 28 =
7.	54 + 24 =	22.	34 + 30 =
8.	73 + 15 =	23.	34 + 29 =
9.	72 + 17 =	24.	38 + 22 =
10.	82 + 17 =	25.	38 + 25 =
11.	90 + 8 =	26.	26 + 38 =
12.	88 + 10 =	27.	27 + 39 =
13.	86 + 12 =	28.	31 + 42 =
14.	84 + 14 =	29.	39 + 42 =
15.	82 + 16 =	30.	39 + 55 =

217 B   Add.   First Half

1.	13 + 3 =	16.	63 + 35 =
2.	5 + 14 =	17.	7 + 26 =
3.	5 + 24 =	18.	26 + 17 =
4.	61 + 22 =	19.	17 + 18 + 18 =
5.	17 + 42 =	20.	9 + 18 + 9 + 18 =
6.	72 + 9 + 8 =	21.	28 + 36 =
7.	53 + 12 + 10 =	22.	37 + 30 =
8.	72 + 8 + 7 =	23.	27 + 39 =
9.	72 + 9 + 8 =	24.	22 + 38 =
10.	82 + 17 =	25.	38 + 25 =
11.	30 + 30 + 30 + 8 =	26.	26 + 38 =
12.	10 + 40 + 48 =	27.	27 + 39 =
13.	86 + 12 =	28.	31 + 21 + 21 =
14.	14 + 84 =	29.	39 + 21 + 21 =
15.	16 + 82 =	30.	39 + 55 =

     Add.

1.	$12 + 3 =$	16.	$63 + 23 =$
2.	$4 + 14 =$	17.	$5 + 26 =$
3.	$4 + 24 =$	18.	$26 + 15 =$
4.	$61 + 24 =$	19.	$15 + 18 + 18 =$
5.	$16 + 43 =$	20.	$9 + 18 + 9 + 16 =$
6.	$72 + 10 + 8 =$	21.	$28 + 34 =$
7.	$53 + 12 + 13 =$	22.	$34 + 10 + 20 =$
8.	$72 + 8 + 8 =$	23.	$27 + 36 =$
9.	$72 + 9 + 8 =$	24.	$22 + 38 =$
10.	$82 + 17 =$	25.	$38 + 25 =$
11.	$30 + 30 + 30 + 8 =$	26.	$26 + 38 =$
12.	$10 + 40 + 48 =$	27.	$27 + 39 =$
13.	$86 + 12 =$	28.	$31 + 21 + 21 =$
14.	$14 + 84 =$	29.	$39 + 21 + 21 =$
15.	$16 + 82 =$	30.	$39 + 55 =$

218 A		Add.	First Half

1.	19 + 4 =	11.	171 + 14 =
2.	29 + 4 =	12.	162 + 9 =
3.	39 + 3 =	13.	287 + 5 =
4.	49 + 4 =	14.	287 + 50 =
5.	59 + 6 =	15.	235 + 70 =
6.	58 + 6 =	16.	235 + 75 =
7.	58 + 7 =	17.	236 + 70 =
8.	68 + 7 =	18.	236 + 75 =
9.	68 + 8 =	19.	261 + 40 =
10.	70 + 14 =	20.	261 + 50 =

       Add.   Second Half

1.	$19 + 5 =$	11.	$171 + 15 =$
2.	$29 + 5 =$	12.	$143 + 9 =$
3.	$39 + 5 =$	13.	$285 + 5 =$
4.	$49 + 5 =$	14.	$285 + 50 =$
5.	$59 + 5 =$	15.	$235 + 70 =$
6.	$58 + 5 =$	16.	$235 + 75 =$
7.	$58 + 7 =$	17.	$236 + 70 =$
8.	$68 + 7 =$	18.	$236 + 75 =$
9.	$68 + 8 =$	19.	$261 + 40 =$
10.	$70 + 14 =$	20.	$261 + 50 =$

218 B   Add.   First Half

1.	19 + 4 =	11.	171 + 14 =
2.	4 + 29 =	12.	152 + 19 =
3.	3 + 39 =	13.	277 + 15 =
4.	4 + 49 =	14.	287 + 50 =
5.	6 + 59 =	15.	245 + 60 =
6.	40 + 6 + 18 =	16.	235 + 75 =
7.	58 + 7 =	17.	236 + 20 + 50 =
8.	7 + 68 =	18.	236 + 75 =
9.	8 + 68 =	19.	261 + 40 =
10.	14 + 70 =	20.	261 + 50 =

1.	19 + 5 =	11.	172 + 14 =
2.	5 + 29 =	12.	133 + 19 =
3.	23 + 21 =	13.	153 + 137 =
4.	5 + 49 =	14.	262 + 73 =
5.	5 + 59 =	15.	215 + 60 + 30 =
6.	40 + 6 + 17 =	16.	235 + 75 =
7.	58 + 7 =	17.	236 + 20 + 50 =
8.	7 + 68 =	18.	236 + 75 =
9.	8 + 68 =	19.	261 + 40 =
10.	14 + 70 =	20.	261 + 50 =

218 B Add. Second Half

219 A Add. First Half

1.	9 + 1 =		11.	27 + 98 =
2.	4 + 9 =		12.	45 + 99 =
3.	4 + 99 =		13.	93 + 99 =
4.	6 + 99 =		14.	36 + 99 =
5.	6 + 98 =		15.	64 + 97 =
6.	8 + 98 =		16.	87 + 98 =
7.	31 + 98 =		17.	94 + 99 =
8.	33 + 98 =		18.	89 + 99 =
9.	46 + 98 =		19.	98 + 88 =
10.	98 + 46 =		20.	99 + 99 =

219 A		Add.		Second Half

1.	9 + 1 =	11.	37 + 98 =
2.	3 + 9 =	12.	47 + 99 =
3.	3 + 99 =	13.	93 + 99 =
4.	5 + 99 =	14.	36 + 99 =
5.	5 + 98 =	15.	64 + 97 =
6.	7 + 98 =	16.	87 + 98 =
7.	21 + 98 =	17.	94 + 99 =
8.	32 + 98 =	18.	89 + 99 =
9.	36 + 98 =	19.	98 + 88 =
10.	98 + 46 =	20.	99 + 99 =

Math Sprints 2

Add.   First Half

1.	$9 + 1 =$	11.	$27 + 98 =$
2.	$4 + 9 =$	12.	$46 + 98 =$
3.	$5 + 98 =$	13.	$94 + 98 =$
4.	$7 + 98 =$	14.	$37 + 98 =$
5.	$6 + 98 =$	15.	$50 + 14 + 97 =$
6.	$8 + 98 =$	16.	$50 + 37 + 98 =$
7.	$31 + 98 =$	17.	$50 + 44 + 99 =$
8.	$33 + 98 =$	18.	$50 + 39 + 99 =$
9.	$46 + 98 =$	19.	$98 + 50 + 38 =$
10.	$97 + 47 =$	20.	$33 + 33 + 33 + 99 =$

Add.

1.	$9 + 1 =$	11.	$37 + 98 =$
2.	$3 + 9 =$	12.	$48 + 98 =$
3.	$5 + 97 =$	13.	$94 + 98 =$
4.	$6 + 90 + 8 =$	14.	$37 + 98 =$
5.	$6 + 90 + 7 =$	15.	$50 + 14 + 97 =$
6.	$4 + 3 + 98 =$	16.	$50 + 37 + 98 =$
7.	$21 + 98 =$	17.	$50 + 44 + 99 =$
8.	$32 + 98 =$	18.	$50 + 39 + 99 =$
9.	$36 + 98 =$	19.	$98 + 50 + 38 =$
10.	$97 + 47 =$	20.	$33 + 33 + 33 + 99 =$

220 A   Add.   First Half

1.	100 + 99 =	11.	98 + 110 =
2.	160 + 99 =	12.	99 + 23 =
3.	162 + 99 =	13.	99 + 230 =
4.	237 + 99 =	14.	98 + 637 =
5.	237 + 98 =	15.	99 + 496 =
6.	256 + 99 =	16.	98 + 202 =
7.	312 + 98 =	17.	858 + 99 =
8.	205 + 99 =	18.	351 + 98 =
9.	308 + 98 =	19.	98 + 203 =
10.	99 + 414 =	20.	98 + 498 =

220 A   Add.   Second Half

1.	$100 + 98 =$	11.	$98 + 120 =$
2.	$160 + 98 =$	12.	$99 + 36 =$
3.	$161 + 99 =$	13.	$99 + 250 =$
4.	$235 + 99 =$	14.	$98 + 637 =$
5.	$235 + 98 =$	15.	$99 + 496 =$
6.	$256 + 99 =$	16.	$98 + 202 =$
7.	$312 + 98 =$	17.	$858 + 99 =$
8.	$205 + 99 =$	18.	$351 + 98 =$
9.	$308 + 98 =$	19.	$98 + 203 =$
10.	$99 + 414 =$	20.	$98 + 498 =$

Math Sprints 2

Add.

1.	$100 + 99 =$	11.	$98 + 100 + 10 =$
2.	$160 + 99 =$	12.	$99 + 23 =$
3.	$162 + 99 =$	13.	$99 + 130 + 100 =$
4.	$99 + 237 =$	14.	$98 + 537 + 100 =$
5.	$98 + 237 =$	15.	$99 + 496 =$
6.	$99 + 256 =$	16.	$98 + 101 + 101 =$
7.	$98 + 312 =$	17.	$99 + 799 + 59 =$
8.	$99 + 205 =$	18.	$210 + 141 + 98 =$
9.	$98 + 308 =$	19.	$98 + 98 + 105 =$
10.	$99 + 414 =$	20.	$98 + 98 + 98 + 302 =$

220 B		Add.	Second Half
1.	$100 + 98 =$	11.	$98 + 80 + 40 =$
2.	$160 + 98 =$	12.	$99 + 36 =$
3.	$162 + 98 =$	13.	$170 + 179 =$
4.	$99 + 200 + 35 =$	14.	$98 + 537 + 100 =$
5.	$98 + 235 =$	15.	$99 + 496 =$
6.	$79 + 276 =$	16.	$98 + 101 + 101 =$
7.	$78 + 332 =$	17.	$99 + 799 + 59 =$
8.	$99 + 205 =$	18.	$210 + 141 + 98 =$
9.	$98 + 308 =$	19.	$98 + 98 + 105 =$
10.	$99 + 414 =$	20.	$98 + 98 + 98 + 302 =$

221 A Subtract. First Half

1.	$8 - 4 =$	11.	$70 - 9 =$
2.	$78 - 4 =$	12.	$270 - 9 =$
3.	$578 - 4 =$	13.	$306 - 9 =$
4.	$578 - 40 =$	14.	$12 - 8 =$
5.	$578 - 400 =$	15.	$312 - 8 =$
6.	$38 - 2 =$	16.	$670 - 50 =$
7.	$338 - 2 =$	17.	$650 - 100 =$
8.	$338 - 20 =$	18.	$650 - 90 =$
9.	$338 - 200 =$	19.	$730 - 100 =$
10.	$70 - 10 =$	20.	$730 - 90 =$

221 A		Subtract.	Second Half

1.	$9 - 4 =$	11.	$80 - 9 =$
2.	$79 - 4 =$	12.	$290 - 9 =$
3.	$569 - 4 =$	13.	$307 - 9 =$
4.	$569 - 40 =$	14.	$12 - 7 =$
5.	$569 - 400 =$	15.	$312 - 7 =$
6.	$36 - 2 =$	16.	$670 - 40 =$
7.	$336 - 2 =$	17.	$650 - 200 =$
8.	$338 - 20 =$	18.	$650 - 90 =$
9.	$338 - 200 =$	19.	$730 - 100 =$
10.	$70 - 10 =$	20.	$730 - 90 =$

	221 B	Subtract.		First Half
1.	$18 - 14 =$		11.	$70 - 9 =$
2.	$88 - 14 =$		12.	$270 - 9 =$
3.	$678 - 104 =$		13.	$306 - 9 =$
4.	$678 - 140 =$		14.	$112 - 108 =$
5.	$588 - 410 =$		15.	$312 - 8 =$
6.	$138 - 102 =$		16.	$675 - 55 =$
7.	$348 - 12 =$		17.	$640 - 90 =$
8.	$358 - 40 =$		18.	$650 - 90 =$
9.	$358 - 220 =$		19.	$830 - 200 =$
10.	$70 - 10 =$		20.	$830 - 190 =$

221 B		Subtract.	Second Half

1.	$18 - 13 =$	11.	$80 - 9 =$
2.	$88 - 13 =$	12.	$290 - 9 =$
3.	$678 - 113 =$	13.	$306 - 8 =$
4.	$678 - 149 =$	14.	$112 - 107 =$
5.	$588 - 419 =$	15.	$312 - 7 =$
6.	$138 - 104 =$	16.	$675 - 45 =$
7.	$348 - 14 =$	17.	$640 - 190 =$
8.	$358 - 40 =$	18.	$650 - 90 =$
9.	$358 - 220 =$	19.	$830 - 200 =$
10.	$70 - 10 =$	20.	$830 - 190 =$

222 A		Subtract.		First Half
1.	$4 - 3 =$		16.	$47 - 36 =$
2.	$14 - 3 =$		17.	$47 - 37 =$
3.	$24 - 3 =$		18.	$47 - 38 =$
4.	$204 - 3 =$		19.	$57 - 37 =$
5.	$38 - 26 =$		20.	$57 - 38 =$
6.	$380 - 260 =$		21.	$67 - 38 =$
7.	$92 - 80 =$		22.	$86 - 44 =$
8.	$82 - 22 =$		23.	$86 - 46 =$
9.	$79 - 35 =$		24.	$90 - 50 =$
10.	$98 - 76 =$		25.	$90 - 51 =$
11.	$99 - 77 =$		26.	$90 - 60 =$
12.	$99 - 78 =$		27.	$90 - 59 =$
13.	$99 - 88 =$		28.	$22 - 15 =$
14.	$95 - 25 =$		29.	$42 - 35 =$
15.	$95 - 35 =$		30.	$52 - 25 =$

222 A   Subtract.   Second Half

1.	$5 - 4 =$	16.	$57 - 36 =$
2.	$15 - 4 =$	17.	$57 - 37 =$
3.	$25 - 4 =$	18.	$57 - 38 =$
4.	$205 - 4 =$	19.	$67 - 37 =$
5.	$37 - 25 =$	20.	$67 - 38 =$
6.	$320 - 200 =$	21.	$77 - 38 =$
7.	$92 - 70 =$	22.	$87 - 44 =$
8.	$82 - 32 =$	23.	$86 - 46 =$
9.	$79 - 35 =$	24.	$90 - 50 =$
10.	$98 - 76 =$	25.	$90 - 51 =$
11.	$99 - 77 =$	26.	$90 - 60 =$
12.	$99 - 78 =$	27.	$90 - 59 =$
13.	$99 - 88 =$	28.	$22 - 15 =$
14.	$95 - 25 =$	29.	$42 - 35 =$
15.	$95 - 35 =$	30.	$52 - 25 =$

222 B   Subtract.   First Half

1.	$14 - 13 =$	16.	$47 - 36 =$
2.	$24 - 13 =$	17.	$47 - 37 =$
3.	$24 - 3 =$	18.	$47 - 38 =$
4.	$204 - 3 =$	19.	$157 - 137 =$
5.	$138 - 126 =$	20.	$157 - 138 =$
6.	$482 - 362 =$	21.	$67 - 38 =$
7.	$92 - 80 =$	22.	$186 - 144 =$
8.	$182 - 122 =$	23.	$186 - 146 =$
9.	$279 - 235 =$	24.	$90 - 50 =$
10.	$198 - 176 =$	25.	$90 - 51 =$
11.	$199 - 177 =$	26.	$90 - 60 =$
12.	$99 - 78 =$	27.	$90 - 59 =$
13.	$99 - 88 =$	28.	$222 - 215 =$
14.	$195 - 125 =$	29.	$342 - 335 =$
15.	$195 - 135 =$	30.	$852 - 825 =$

222 B		Subtract.		Second Half
1.	14 − 13 =	16.	47 − 26 =	
2.	24 − 13 =	17.	57 − 37 =	
3.	24 − 3 =	18.	47 − 28 =	
4.	204 − 3 =	19.	167 − 137 =	
5.	138 − 126 =	20.	167 − 138 =	
6.	482 − 362 =	21.	77 − 38 =	
7.	102 − 80 =	22.	187 − 144 =	
8.	182 − 132 =	23.	186 − 146 =	
9.	279 − 235 =	24.	90 − 50 =	
10.	198 − 176 =	25.	90 − 51 =	
11.	199 − 177 =	26.	90 − 60 =	
12.	99 − 78 =	27.	90 − 59 =	
13.	99 − 88 =	28.	222 − 215 =	
14.	195 − 125 =	29.	342 − 335 =	
15.	195 − 135 =	30.	852 − 825 =	

223 A	Subtract.	First Half

1.	$100 - 100 =$	11.	$203 - 99 =$
2.	$100 - 99 =$	12.	$304 - 99 =$
3.	$200 - 99 =$	13.	$304 - 98 =$
4.	$300 - 99 =$	14.	$306 - 99 =$
5.	$300 - 98 =$	15.	$407 - 99 =$
6.	$400 - 98 =$	16.	$508 - 99 =$
7.	$500 - 99 =$	17.	$350 - 99 =$
8.	$700 - 98 =$	18.	$253 - 99 =$
9.	$701 - 98 =$	19.	$717 - 99 =$
10.	$801 - 99 =$	20.	$423 - 98 =$

223 A   Subtract.   Second Half

1.	$200 - 200 =$	11.	$302 - 99 =$
2.	$200 - 199 =$	12.	$305 - 99 =$
3.	$200 - 99 =$	13.	$305 - 98 =$
4.	$300 - 99 =$	14.	$307 - 99 =$
5.	$200 - 98 =$	15.	$408 - 99 =$
6.	$400 - 98 =$	16.	$507 - 99 =$
7.	$500 - 97 =$	17.	$450 - 99 =$
8.	$600 - 98 =$	18.	$264 - 99 =$
9.	$701 - 98 =$	19.	$716 - 99 =$
10.	$801 - 99 =$	20.	$413 - 99 =$

223 B   Subtract.   First Half

1.	$100 - 100 =$	11.	$203 - 99 =$
2.	$100 - 99 =$	12.	$404 - 199 =$
3.	$200 - 99 =$	13.	$404 - 198 =$
4.	$300 - 99 =$	14.	$406 - 199 =$
5.	$300 - 98 =$	15.	$407 - 99 =$
6.	$400 - 98 =$	16.	$608 - 199 =$
7.	$500 - 99 =$	17.	$350 - 99 =$
8.	$700 - 98 =$	18.	$252 - 98 =$
9.	$701 - 98 =$	19.	$716 - 98 =$
10.	$801 - 99 =$	20.	$523 - 198 =$

223 B		Subtract.		Second Half

1.	$200 - 200 =$	11.	$302 - 99 =$	
2.	$200 - 199 =$	12.	$405 - 199 =$	
3.	$300 - 199 =$	13.	$405 - 198 =$	
4.	$400 - 199 =$	14.	$407 - 199 =$	
5.	$400 - 298 =$	15.	$508 - 199 =$	
6.	$500 - 198 =$	16.	$607 - 199 =$	
7.	$500 - 91 - 6 =$	17.	$550 - 199 =$	
8.	$800 - 298 =$	18.	$264 - 99 =$	
9.	$701 - 98 =$	19.	$816 - 199 =$	
10.	$801 - 99 =$	20.	$523 - 209 =$	

224 A   Count by threes and fill in the blanks.   First Half

1.	3, 6, 9, _____	16.	24, _____, 30
2.	9, 12, 15, _____	17.	3, _____, 9
3.	21, 24, _____	18.	3, 6, _____
4.	6, 9, _____	19.	9, _____, 15
5.	15, 18, _____	20.	15, 18, _____
6.	30, 27, _____	21.	21, _____, 15
7.	18, 15, _____	22.	27, _____, 21
8.	24, 21, _____	23.	30, _____, 24
9.	30, _____, 24	24.	15, _____, 9
10.	9, _____, 3	25.	18, _____, 12
11.	18, 15, _____	26.	6, _____, 12
12.	12, _____, 18	27.	3, 6, _____
13.	21, _____, 27	28.	12, _____, 18
14.	6, _____, 12	29.	21, 24, _____
15.	15, _____, 21	30.	30, _____, 24

224 A Count by threes and fill in the blanks. **Second Half**

1.	0, 3, 6, _____	16.	21, _____, 27
2.	6, 9, 12, _____	17.	0, 3, _____, 9
3.	15, 18, 21, _____	18.	6, _____, 12
4.	6, 9, _____	19.	15, _____, 9
5.	12, 15, _____	20.	12, 15, _____
6.	33, 30, _____	21.	21, _____, 15
7.	21, 18, _____	22.	27, _____, 21
8.	24, 21, _____	23.	30, _____, 24
9.	30, _____, 24	24.	15, _____, 9
10.	9, _____, 3	25.	18, _____, 12
11.	18, 15, _____	26.	6, _____, 12
12.	12, _____, 18	27.	3, 6, _____
13.	21, _____, 27	28.	12, _____, 18
14.	6, _____, 12	29.	21, 24, _____
15.	15, _____, 21	30.	30, _____, 24

Fill in the blanks.

1.	3, 6, 9, _____	16.	30, _____, 24, 21
2.	9, 12, 15, _____	17.	12, 9, _____, 3
3.	30, _____, 24, 21	18.	15, 12, _____, 6
4.	4, 8, _____, 16	19.	8, _____, 16, 20
5.	27, 24, _____, 18	20.	7, 14, _____, 28
6.	16, 20, _____, 28	21.	6, 12, _____, 24
7.	9, _____, 15, 18	22.	21, _____, 27, 30
8.	21, _____, 15, 12	23.	30, _____, 24, 21
9.	30, _____, 24, 21	24.	6, _____, 18, 24
10.	9, _____, 3, 0	25.	25, 20, _____, 10
11.	36, 24, _____, 0	26.	_____, 18, 27, 36
12.	5, 10, _____, 20	27.	3, 6, _____, 12
13.	30, 27, _____, 21	28.	21, 18, _____, 12
14.	27, 18, _____, 0	29.	33, 30, _____, 24
15.	15, _____, 21, 24	30.	9, 18, _____, 36

Fill in the blanks.

1.	3, 6, _____, 12	16.	30, 27, _____, 21
2.	9, 12, _____	17.	12, 9, _____, 3
3.	30, 27, _____, 21	18.	15, 12, _____, 6
4.	4, 8, _____, 16	19.	20, 16, _____
5.	27, 24, 21, _____	20.	24, 21, _____
6.	24, _____, 30	21.	6, 12, _____, 24
7.	9, 12, _____, 18	22.	27, _____, 21
8.	21, _____, 15, 12	23.	30, _____, 24, 21
9.	30, _____, 24, 21	24.	6, _____, 18, 24
10.	9, _____, 3, 0	25.	25, 20, _____, 10
11.	36, 24, _____, 0	26.	_____, 18, 27, 36
12.	5, 10, _____, 20	27.	3, 6, _____, 12
13.	30, 27, _____, 21	28.	21, 18, _____, 12
14.	27, 18, _____, 0	29.	33, 30, _____, 24
15.	15, _____, 21, 24	30.	9, 18, _____, 36

225 A Multiply. First Half

1.	$4 \times 1 =$	11.	$9 \times 4 =$
2.	$4 \times 2 =$	12.	$4 \times 10 =$
3.	$4 \times 3 =$	13.	$4 \times 0 =$
4.	$1 \times 4 =$	14.	$4 \times 9 =$
5.	$2 \times 4 =$	15.	$8 \times 4 =$
6.	$3 \times 4 =$	16.	$7 \times 4 =$
7.	$5 \times 4 =$	17.	$4 \times 6 =$
8.	$6 \times 4 =$	18.	$10 \times 4 =$
9.	$4 \times 7 =$	19.	$4 \times 9 =$
10.	$4 \times 8 =$	20.	$4 \times 4 =$

225 A   Multiply.   Second Half

1.	4 x 0 =	11.	7 x 4 =
2.	4 x 1 =	12.	4 x 10 =
3.	4 x 2 =	13.	4 x 0 =
4.	2 x 4 =	14.	4 x 3 =
5.	3 x 4 =	15.	8 x 4 =
6.	4 x 4 =	16.	7 x 4 =
7.	6 x 4 =	17.	4 x 6 =
8.	7 x 4 =	18.	10 x 4 =
9.	4 x 8 =	19.	4 x 9 =
10.	9 x 4 =	20.	4 x 4 =

225 B   Multiply or divide.   First Half

1.	$24 \div 6 =$	11.	$9 \times 4 =$
2.	$32 \div 4 =$	12.	$4 \times 10 =$
3.	$4 \times 3 =$	13.	$0 \times 4 =$
4.	$16 \div 4 =$	14.	$2 \times 9 \times 2 =$
5.	$32 \div 4 =$	15.	$2 \times 2 \times 2 \times 2 \times 2 =$
6.	$2 \times 2 \times 3 =$	16.	$7 \times 4 =$
7.	$5 \times 4 =$	17.	$2 \times 12 =$
8.	$2 \times 2 \times 2 \times 3 =$	18.	$5 \times 2 \times 4 =$
9.	$2 \times 7 \times 2 =$	19.	$2 \times 2 \times 3 \times 3 =$
10.	$4 \times 8 =$	20.	$4 \times 4 =$

225 B   Multiply or divide.   Second Half

1.	$4 \times 0 =$	11.	$7 \times 4 =$
2.	$32 \div 8 =$	12.	$5 \times 4 \times 2 =$
3.	$32 \div 4 =$	13.	$0 \times 4 =$
4.	$4 \times 2 =$	14.	$2 \times 3 \times 2 =$
5.	$48 \div 4 =$	15.	$2 \times 8 \times 2 =$
6.	$2 \times 2 \times 2 \times 2 =$	16.	$7 \times 4 =$
7.	$6 \times 4 =$	17.	$2 \times 3 \times 4 =$
8.	$56 \div 2 =$	18.	$5 \times 2 \times 4 =$
9.	$2 \times 2 \times 2 \times 2 \times 2 =$	19.	$2 \times 2 \times 3 \times 3 =$
10.	$4 \times 9 =$	20.	$64 \div 4 =$

226 A		Divide.		First Half

1.	$4 \div 4 =$	16.	$9 \div 3 =$
2.	$3 \div 3 =$	17.	$21 \div 3 =$
3.	$2 \div 2 =$	18.	$15 \div 3 =$
4.	$4 \div 1 =$	19.	$27 \div 3 =$
5.	$3 \div 1 =$	20.	$18 \div 3 =$
6.	$2 \div 1 =$	21.	$30 \div 3 =$
7.	$8 \div 4 =$	22.	$20 \div 2 =$
8.	$16 \div 4 =$	23.	$14 \div 2 =$
9.	$12 \div 4 =$	24.	$16 \div 2 =$
10.	$12 \div 3 =$	25.	$8 \div 2 =$
11.	$20 \div 4 =$	26.	$12 \div 2 =$
12.	$28 \div 4 =$	27.	$4 \div 2 =$
13.	$40 \div 4 =$	28.	$6 \div 2 =$
14.	$32 \div 4 =$	29.	$36 \div 4 =$
15.	$36 \div 4 =$	30.	$18 \div 2 =$

	226 A		Divide.		Second Half

1.	$2 \div 2 =$	16.	$12 \div 4 =$
2.	$4 \div 4 =$	17.	$28 \div 4 =$
3.	$3 \div 3 =$	18.	$15 \div 3 =$
4.	$4 \div 1 =$	19.	$27 \div 3 =$
5.	$3 \div 1 =$	20.	$18 \div 3 =$
6.	$2 \div 1 =$	21.	$30 \div 3 =$
7.	$10 \div 5 =$	22.	$20 \div 2 =$
8.	$8 \div 2 =$	23.	$14 \div 2 =$
9.	$12 \div 4 =$	24.	$16 \div 2 =$
10.	$12 \div 3 =$	25.	$8 \div 2 =$
11.	$20 \div 4 =$	26.	$12 \div 2 =$
12.	$28 \div 4 =$	27.	$4 \div 2 =$
13.	$40 \div 4 =$	28.	$6 \div 2 =$
14.	$32 \div 4 =$	29.	$36 \div 4 =$
15.	$36 \div 4 =$	30.	$18 \div 2 =$

Fill in the blanks.

1.	$4 \times \underline{\hspace{2cm}} = 4$	16.	$\underline{\hspace{2cm}} \times 8 = 24$
2.	$\underline{\hspace{2cm}} \times 3 = 3$	17.	$4 \times \underline{\hspace{2cm}} = 28$
3.	$2 \times \underline{\hspace{2cm}} = 2$	18.	$3 \times \underline{\hspace{2cm}} = 15$
4.	$4 \times \underline{\hspace{2cm}} = 16$	19.	$\underline{\hspace{2cm}} \times 4 = 36$
5.	$1 \times \underline{\hspace{2cm}} = 3$	20.	$3 \times \underline{\hspace{2cm}} = 18$
6.	$\underline{\hspace{2cm}} \times 1 = 2$	21.	$\underline{\hspace{2cm}} \times 3 = 30$
7.	$\underline{\hspace{2cm}} \times 4 = 8$	22.	$2 \times \underline{\hspace{2cm}} = 20$
8.	$5 \times \underline{\hspace{2cm}} = 20$	23.	$\underline{\hspace{2cm}} \times 3 = 21$
9.	$\underline{\hspace{2cm}} \times 9 = 27$	24.	$\underline{\hspace{2cm}} \times 2 = 16$
10.	$6 \times \underline{\hspace{2cm}} = 24$	25.	$\underline{\hspace{2cm}} \times 4 = 16$
11.	$4 \times \underline{\hspace{2cm}} = 20$	26.	$4 \times \underline{\hspace{2cm}} = 24$
12.	$\underline{\hspace{2cm}} \times 3 = 21$	27.	$\underline{\hspace{2cm}} \times 9 = 18$
13.	$3 \times \underline{\hspace{2cm}} = 30$	28.	$6 \times \underline{\hspace{2cm}} = 18$
14.	$4 \times \underline{\hspace{2cm}} = 32$	29.	$\underline{\hspace{2cm}} \times 4 = 36$
15.	$\underline{\hspace{2cm}} \times 2 = 18$	30.	$3 \times \underline{\hspace{2cm}} = 27$

Fill in the blanks.

1.	$4 \times \underline{\hspace{2cm}} = 4$	16.	$\underline{\hspace{2cm}} \times 8 = 24$
2.	$\underline{\hspace{2cm}} \times 3 = 3$	17.	$4 \times \underline{\hspace{2cm}} = 28$
3.	$2 \times \underline{\hspace{2cm}} = 2$	18.	$3 \times \underline{\hspace{2cm}} = 15$
4.	$4 \times \underline{\hspace{2cm}} = 16$	19.	$\underline{\hspace{2cm}} \times 4 = 36$
5.	$1 \times \underline{\hspace{2cm}} = 3$	20.	$3 \times \underline{\hspace{2cm}} = 18$
6.	$\underline{\hspace{2cm}} \times 1 = 2$	21.	$\underline{\hspace{2cm}} \times 3 = 30$
7.	$\underline{\hspace{2cm}} \times 4 = 8$	22.	$2 \times \underline{\hspace{2cm}} = 20$
8.	$5 \times \underline{\hspace{2cm}} = 20$	23.	$\underline{\hspace{2cm}} \times 3 = 21$
9.	$\underline{\hspace{2cm}} \times 9 = 27$	24.	$\underline{\hspace{2cm}} \times 2 = 16$
10.	$6 \times \underline{\hspace{2cm}} = 24$	25.	$\underline{\hspace{2cm}} \times 4 = 16$
11.	$4 \times \underline{\hspace{2cm}} = 20$	26.	$4 \times \underline{\hspace{2cm}} = 24$
12.	$\underline{\hspace{2cm}} \times 3 = 21$	27.	$\underline{\hspace{2cm}} \times 9 = 18$
13.	$3 \times \underline{\hspace{2cm}} = 30$	28.	$6 \times \underline{\hspace{2cm}} = 18$
14.	$4 \times \underline{\hspace{2cm}} = 32$	29.	$\underline{\hspace{2cm}} \times 4 = 36$
15.	$\underline{\hspace{2cm}} \times 2 = 18$	30.	$3 \times \underline{\hspace{2cm}} = 27$

	227 A		Multiply.		First Half

1.	$2 \times 1 =$	16.	$2 \times 4 =$
2.	$2 \times 2 =$	17.	$4 \times 4 =$
3.	$4 \times 1 =$	18.	$2 \times 5 =$
4.	$2 \times 6 =$	19.	$4 \times 5 =$
5.	$4 \times 3 =$	20.	$2 \times 6 =$
6.	$2 \times 8 =$	21.	$4 \times 6 =$
7.	$4 \times 4 =$	22.	$2 \times 7 =$
8.	$1 \times 2 =$	23.	$4 \times 7 =$
9.	$1 \times 4 =$	24.	$2 \times 8 =$
10.	$6 \times 2 =$	25.	$4 \times 8 =$
11.	$3 \times 4 =$	26.	$2 \times 9 =$
12.	$8 \times 2 =$	27.	$4 \times 9 =$
13.	$4 \times 4 =$	28.	$2 \times 10 =$
14.	$2 \times 3 =$	29.	$4 \times 10 =$
15.	$4 \times 3 =$	30.	$2 \times 100 =$

227 A   Multiply.   Second Half

1.	2 x 0 =	16.	2 x 4 =
2.	2 x 1 =	17.	4 x 4 =
3.	4 x 0 =	18.	2 x 5 =
4.	2 x 6 =	19.	4 x 5 =
5.	4 x 4 =	20.	2 x 6 =
6.	2 x 8 =	21.	4 x 6 =
7.	4 x 4 =	22.	2 x 7 =
8.	2 x 2 =	23.	4 x 7 =
9.	2 x 4 =	24.	2 x 8 =
10.	6 x 2 =	25.	4 x 8 =
11.	3 x 3 =	26.	2 x 9 =
12.	7 x 2 =	27.	4 x 9 =
13.	4 x 7 =	28.	2 x 10 =
14.	2 x 6 =	29.	4 x 10 =
15.	4 x 3 =	30.	2 x 100 =

227 B   Divide.   First Half

1.	$20 \div 10 =$	16.	$32 \div 4 =$
2.	$20 \div 5 =$	17.	$32 \div 2 =$
3.	$16 \div 4 =$	18.	$100 \div 10 =$
4.	$24 \div 2 =$	19.	$100 \div 5 =$
5.	$12 \div 1 =$	20.	$48 \div 4 =$
6.	$160 \div 10 =$	21.	$48 \div 2 =$
7.	$32 \div 2 =$	22.	$28 \div 2 =$
8.	$8 \div 4 =$	23.	$280 \div 10 =$
9.	$8 \div 2 =$	24.	$160 \div 10 =$
10.	$36 \div 3 =$	25.	$320 \div 10 =$
11.	$120 \div 10 =$	26.	$180 \div 10 =$
12.	$16 \div 1 =$	27.	$360 \div 10 =$
13.	$32 \div 2 =$	28.	$100 \div 5 =$
14.	$24 \div 4 =$	29.	$200 \div 5 =$
15.	$24 \div 2 =$	30.	$400 \div 2 =$

227 B		Multiply or divide.		Second Half
1.	20 x 0 =	16.	32 ÷ 4 =
2.	20 ÷ 10 =	17.	32 ÷ 2 =
3.	16 x 0 =	18.	100 ÷ 10 =
4.	36 ÷ 3 =	19.	100 ÷ 5 =
5.	16 ÷ 1 =	20.	48 ÷ 4 =
6.	64 ÷ 4 =	21.	48 ÷ 2 =
7.	48 ÷ 3 =	22.	28 ÷ 2 =
8.	8 ÷ 2 =	23.	280 ÷ 10 =
9.	40 ÷ 5 =	24.	160 ÷ 10 =
10.	48 ÷ 4 =	25.	320 ÷ 10 =
11.	99 ÷ 11 =	26.	180 ÷ 10 =
12.	14 ÷ 1 =	27.	360 ÷ 10 =
13.	56 ÷ 2 =	28.	100 ÷ 5 =
14.	24 ÷ 2 =	29.	200 ÷ 5 =
15.	60 ÷ 5 =	30.	400 ÷ 2 =

228 A   Subtract.   First Half

1.	$1.00 − $1.00 = _____ ¢	16.	$1.00 − 15¢ = _____ ¢
2.	$0.45 − 10¢ = _____ ¢	17.	$1.00 − 35¢ = _____ ¢
3.	$1.00 − 20¢ = _____ ¢	18.	$1.00 − 45¢ = _____ ¢
4.	$1.00 − 25¢ = _____ ¢	19.	$1.00 − 65¢ = _____ ¢
5.	$1.00 − 50¢ = _____ ¢	20.	$1.00 − 85¢ = _____ ¢
6.	$1.00 − 75¢ = _____ ¢	21.	$1.00 − 95¢ = _____ ¢
7.	$1.00 − 5¢ = _____ ¢	22.	$1.00 − 75¢ = _____ ¢
8.	$1.00 − 30¢ = _____ ¢	23.	$1.00 − 5¢ = _____ ¢
9.	$1.00 − 60¢ = _____ ¢	24.	$1.00 − 95¢ = _____ ¢
10.	$1.00 − 40¢ = _____ ¢	25.	$1.00 − 10¢ = _____ ¢
11.	$1.00 − 70¢ = _____ ¢	26.	$1.00 − 90¢ = _____ ¢
12.	$1.00 − 90¢ = _____ ¢	27.	$1.00 − $1.00 = _____ ¢
13.	$1.00 − 80¢ = _____ ¢	28.	$1.00 − 15¢ = _____ ¢
14.	$1.00 − 85¢ = _____ ¢	29.	$1.00 − 35¢ = _____ ¢
15.	$1.00 − 75¢ = _____ ¢	30.	$1.00 − 45¢ = _____ ¢

228 A Subtract. Second Half

1.	$2.00 − $2.00 = _____ ¢	16.	$1.00 − 15¢ = _____ ¢
2.	$1.00 − 10¢ = _____ ¢	17.	$1.00 − 35¢ = _____ ¢
3.	$1.00 − 20¢ = _____ ¢	18.	$1.00 − 45¢ = _____ ¢
4.	$1.00 − 25¢= _____ ¢	19.	$1.00 − 65¢ = _____ ¢
5.	$1.00 − 50¢ = _____ ¢	20.	$1.00 − 85¢ = _____ ¢
6.	$1.00 − 75¢ = _____ ¢	21.	$1.00 − 95¢ = _____ ¢
7.	$2.00 − $1.05 = _____ ¢	22.	$1.00 − 75¢ = _____ ¢
8.	$1.00 − 30¢ = _____ ¢	23.	$1.00 − 5¢ = _____ ¢
9.	$1.00 − 60¢= _____ ¢	24.	$1.00 − 95¢ = _____ ¢
10.	$1.00 − 40¢= _____ ¢	25.	$1.00 − 10¢ = _____ ¢
11.	$1.00 − 70¢ = _____ ¢	26.	$1.00 − 90¢ = _____ ¢
12.	$1.00 − 90 ¢ = _____ ¢	27.	$1.00 − $1.00 = _____ ¢
13.	$1.00 − 80¢ = _____ ¢	28.	$1.00 − 15¢ = _____ ¢
14.	$1.00 − 85¢ = _____ ¢	29.	$1.00 − 35¢ = _____ ¢
15.	$1.00 − 75¢ = _____ ¢	30.	$1.00 − 45¢ = _____ ¢

Math Sprints 2

Fill in the blanks.

1.	$1.00 – $1.00 = _____ ¢	16.	$2.10 – $1.25 = _____ ¢
2.	$1.00 – 65¢ = _____ ¢	17.	_____ ¢ + $1.35 = $2.00
3.	25¢ + 25¢ + 25¢ + 5¢ = _____ ¢	18.	_____ ¢ + $ 1.15 = $ 1.70
4.	$2.10 – $1.35 = _____ ¢	19.	$1.05 – 70¢ = _____ ¢
5.	$5.00 – $4.50 = _____ ¢	20.	$1.45 + _____ ¢ = $1.60
6.	4 nickels and five pennies = _____ ¢	21.	_____ ¢ + 81¢ = 86¢
7.	15 nickels and two dimes = _____ ¢	22.	2 nickels, 1 dime, 5 pennies = _____ ¢
8.	$2.40 – $1.70 = _____ ¢	23.	Nineteen nickels = _____ ¢
9.	$ 1.10 – 70¢ = _____ ¢	24.	$2.10 – $2.05 = = _____ ¢
10.	$1.50 – 90¢ = _____ ¢	25.	3 dimes, 8 nickels, 20 pennies = _____ ¢
11.	$2.05 – $1.75 = _____ ¢	26.	Ten pennies = _____ ¢
12.	$1.00 ÷ 10 = _____ ¢	27.	$4.50 – $4.50 = _____ ¢
13.	1 dime, 1 nickel, 5 pennies = _____ ¢	28.	15 nickels and one dime = _____ ¢
14.	$12.00 -$11.85 = _____ ¢	29.	$3.10 – $2.45 = _____ ¢
15.	$14.10 – $13.85 = _____ ¢	30.	$1.00 – 45¢ = _____ ¢

228 B Fill in the blanks. Second Half

1.	$1.00 − $1.00 = _____ ¢	16.	$2.10 − $1.25 = _____ ¢
2.	$1.00 − 10¢ = _____ ¢	17.	_____ ¢ + $1.35 = $2.00
3.	25¢ + 25¢ + 25¢ + 5¢ = _____ ¢	18.	_____ ¢ + $ 1.15 = $ 1.70
4.	$2.10 − $1.35 = _____ ¢	19.	$1.05 − 70¢ = _____ ¢
5.	$5.00 − $4.50 = _____ ¢	20.	$1.45 + _____ ¢ = $1.60
6.	4 nickels and five pennies = _____ ¢	21.	_____ ¢ + 81¢ = 86¢
7.	15 nickels and two dimes = _____ ¢	22.	2 nickels, 1 dime, 5 pennies = _____ ¢
8.	$2.40 − $1.70 = _____ ¢	23.	Nineteen nickels = _____ ¢
9.	$ 1.10 − 70¢ = _____ ¢	24.	$2.10 − $2.05 = _____ ¢
10.	$1.50 − 90¢ = _____ ¢	25.	3 dimes, 8 nickels, 20 pennies = _____ ¢
11.	$2.05 − $1.75 = _____ ¢	26.	Ten pennies = _____ ¢
12.	$1.00 ÷ 10 = _____ ¢	27.	$4.50 − $4.50 = _____ ¢
13.	One dime, one nickel, five pennies =_____ ¢	28.	13 nickels and two dimes = _____ ¢
14.	$12.00 -$11.85 = _____ ¢	29.	$3.10 − $2.45 = _____ ¢
15.	$14.10 − $13.85 = _____ ¢	30.	$1.00 − 45¢ = _____ ¢

Math Sprints 2

Add or subtract.

1.	$2.75 + $2 = $_____	11.	$8.50 − $3 = $_____
2.	$1.50 + $3 = $_____	12.	$7.95 − $2 = $_____
3.	$6.25 + $4 = $_____	13.	$9.15 − $4 = $_____
4.	$4.85 + $3 = $_____	14.	$6.45 − $6 = _____ ¢
5.	50¢ + 50¢ = $_____	15.	$1 − 30¢ = _____ ¢
6.	$1.50 + 50¢ = $_____	16.	$4 − 30¢ = $_____
7.	75¢ + 25¢ = $_____	17.	$6 − 75¢ = $_____
8.	$5.75 + 25¢ = $_____	18.	$2 − 50¢ = $_____
9.	85¢ + 15¢ = $_____	19.	$5 − 45¢ = $_____
10.	$4.85 + 15¢ = $_____	20.	$8 − 5¢ = $_____

Math Sprints 2

Add or subtract.

1.	$1.75 + $1 = $_____	11.	$5.50 – $3 = $_____
2.	$2.50 + $1 = $_____	12.	$6.95 – $2 = $_____
3.	$3.25 + $3 = $_____	13.	$6.15 – $4 = $_____
4.	$3.85 + $4 = $_____	14.	$6.45 – $4 = $_____
5.	50¢ + 50¢ = $_____	15.	$1 – 30¢ = _____ ¢
6.	$1.50 + 50¢ = $_____	16.	$5 – 30¢ = $_____
7.	75 ¢ + 25¢ = $_____	17.	$5 – 75¢ = $_____
8.	$5.50 + 25¢ = $_____	18.	$2 – 50¢ = $_____
9.	85 ¢ + 15¢ = $_____	19.	$5 – 45¢ = $_____
10.	$3.85 + 15¢ = $_____	20.	$8 – 5¢ = $_____

Math Sprints 2

1.	$\$2.75 + \$2 = \$\underline{\hspace{2cm}}$	11.	$\$8.30 - \$2.80 = \$\underline{\hspace{2cm}}$
2.	$\$1.50 + \$3 = \$\underline{\hspace{2cm}}$	12.	$\$7.85 - \$1.90 = \$\underline{\hspace{2cm}}$
3.	$\$6.15 + \$4.10 = \$\underline{\hspace{2cm}}$	13.	$\$9.05 - \$3.90 = \$\underline{\hspace{2cm}}$
4.	$\$4.45 + \$3.40 = \$\underline{\hspace{2cm}}$	14.	$\$6.35 - \$5.90 = \underline{\hspace{2cm}}¢$
5.	$50¢ + 50¢ = \$\underline{\hspace{2cm}}$	15.	$\$2 - \$1.30 = \underline{\hspace{2cm}}¢$
6.	$\$1.25 + 75¢ = \$\underline{\hspace{2cm}}$	16.	$\$4.35 - 65¢ = \$\underline{\hspace{2cm}}$
7.	$85¢ + 15¢ = \$\underline{\hspace{2cm}}$	17.	$\$6.15 - 90¢ = \$\underline{\hspace{2cm}}$
8.	$\$5.15 + 25¢ + 60¢ = \$\underline{\hspace{2cm}}$	18.	$\$2.15 - 65¢ = \$\underline{\hspace{2cm}}$
9.	$45¢ + 15¢ + 40¢ = \$\underline{\hspace{2cm}}$	19.	$\$5.15 - 60¢ = \$\underline{\hspace{2cm}}$
10.	$\$4.65 + 15¢ + 20¢ = \$\underline{\hspace{2cm}}$	20.	$\$8.05 - 10¢ = \$\underline{\hspace{2cm}}$

229 B   Add or subtract.   Second Half

1.	$1.75 + $1 = $_____	11.	$5.10 − $2.60 = $_____
2.	$2.50 + $1 = $_____	12.	$7.15 − $2 .20 = $_____
3.	$3.25 + $1 + $2 = $_____	13.	$6.15 − $4 = $_____
4.	$3.85 + $2 + $2 = $_____	14.	$6.45 − $4 = $_____
5.	50¢ + 40¢ + 10¢= $_____	15.	$1 − 30¢ = _____¢
6.	$1.10 + 90¢ = $_____	16.	$5 − 30¢ = $_____
7.	75¢ + 25¢ = $_____	17.	$5 − 75¢ = $_____
8.	$4.50 + 25¢ + $1= $_____	18.	$2 − 50¢ = $_____
9.	65¢ + 25¢ + 10¢= $_____	19.	$5.10 − 55¢ = $_____
10.	$2.85 + $1.15 = $_____	20.	$8 − 5¢ = $_____

Answers

201 A & B			First Half
1.	18	11.	830
2.	56	12.	833
3.	123	13.	762
4.	255	14.	702
5.	480	15.	760
6.	173	16.	555
7.	160	17.	505
8.	379	18.	999
9.	564	19.	909
10.	803	20.	990

202 A & B			First Half
1.	<	16.	>
2.	<	17.	>
3.	>	18.	<
4.	<	19.	<
5.	>	20.	<
6.	<	21.	<
7.	<	22.	<
8.	>	23.	>
9.	>	24.	<
10.	<	25.	<
11.	>	26.	<
12.	>	27.	<
13.	>	28.	>
14.	>	29.	>
15.	<	30.	<

203 A & B			First Half
1.	4	16.	500
2.	7	17.	1000
3.	30	18.	405
4.	69	19.	140
5.	125	20.	240
6.	300	21.	130
7.	400	22.	699
8.	40	23.	887
9.	60	24.	210
10.	140	25.	460
11.	260	26.	644
12.	370	27.	401
13.	520	28.	710
14.	650	29.	300
15.	300	30.	990

201 A & B			Second Half
1.	16	11.	830
2.	46	12.	833
3.	114	13.	762
4.	268	14.	702
5.	320	15.	760
6.	374	16.	555
7.	250	17.	606
8.	379	18.	888
9.	564	19.	808
10.	803	20.	880

202 A & B			Second Half
1.	<	16.	>
2.	<	17.	>
3.	>	18.	<
4.	<	19.	<
5.	>	20.	<
6.	<	21.	<
7.	<	22.	<
8.	>	23.	>
9.	>	24.	<
10.	<	25.	<
11.	>	26.	<
12.	>	27.	<
13.	>	28.	>
14.	>	29.	>
15.	<	30.	<

204 A & B			Second Half
1.	5	16.	205
2.	8	17.	1000
3.	20	18.	402
4.	39	19.	140
5.	116	20.	310
6.	273	21.	130
7.	410	22.	699
8.	80	23.	887
9.	80	24.	210
10.	140	25.	460
11.	260	26.	643
12.	370	27.	401
13.	520	28.	710
14.	650	29.	300
15.	300	30.	990

Answers

204 A & B			First Half
1.	4	16.	82
2.	5	17.	84
3.	8	18.	94
4.	10	19.	92
5.	20	20.	95
6.	30	21.	100
7.	31	22.	102
8.	32	23.	102
9.	22	24.	200
10.	40	25.	400
11.	50	26.	501
12.	60	27.	525
13.	70	28.	528
14.	71	29.	538
15.	72	30.	500

205 A & B			First Half
1.	4	16.	33
2.	6	17.	23
3.	8	18.	13
4.	6	19.	2
5.	9	20.	10
6.	22	21.	15
7.	20	22.	25
8.	22	23.	22
9.	12	24.	21
10.	10	25.	30
11.	20	26.	28
12.	10	27.	0
13.	10	28.	25
14.	14	29.	50
15.	24	30.	75

206 A & B			First Half
1.	2	11.	510
2.	12	12.	110
3.	10	13.	11
4.	100	14.	312
5.	111	15.	100
6.	113	16.	188
7.	231	17.	100
8.	130	18.	220
9.	101	19.	181
10.	208	20.	224

204 A & B			Second Half
1.	3	16.	72
2.	4	17.	73
3.	5	18.	84
4.	9	19.	92
5.	19	20.	95
6.	29	21.	100
7.	30	22.	102
8.	31	23.	102
9.	22	24.	200
10.	40	25.	400
11.	50	26.	601
12.	60	27.	525
13.	60	28.	528
14.	61	29.	437
15.	62	30.	400

205 A & B			Second Half
1.	3	16.	23
2.	6	17.	13
3.	7	18.	3
4.	6	19.	2
5.	8	20.	10
6.	22	21.	15
7.	10	22.	25
8.	12	23.	22
9.	2	24.	21
10.	0	25.	30
11.	30	26.	28
12.	10	27.	0
13.	10	28.	25
14.	14	29.	50
15.	24	30.	75

206 A & B			Second Half
1.	3	11.	420
2.	13	12.	110
3.	11	13.	20
4.	100	14.	222
5.	122	15.	100
6.	113	16.	188
7.	231	17.	100
8.	130	18.	220
9.	101	19.	181
10.	208	20.	224

Answers

207 A & B			First Half
1.	10	16.	300
2.	11	17.	500
3.	21	18.	510
4.	31	19.	609
5.	41	20.	610
6.	51	21.	620
7.	151	22.	1000
8.	251	23.	1000
9.	252	24.	1000
10.	352	25.	999
11.	452	26.	1000
12.	70	27.	601
13.	170	28.	701
14.	270	29.	711
15.	270	30.	777

208 A & B			First Half
1.	4	11.	291
2.	14	12.	281
3.	13	13.	278
4.	113	14.	23
5.	9	15.	23
6.	109	16.	58
7.	99	17.	48
8.	195	18.	28
9.	185	19.	128
10.	165	20.	24

209 A & B			First Half
1.	100	11.	15
2.	200	12.	25
3.	300	13.	15
4.	10	14.	13
5.	31	15.	27
6.	8	16.	16
7.	18	17.	33
8.	18	18.	33
9.	30	19.	39
10.	25	20.	75

207 A & B			Second Half
1.	11	16.	500
2.	12	17.	600
3.	22	18.	510
4.	32	19.	605
5.	42	20.	810
6.	52	21.	820
7.	152	22.	1000
8.	252	23.	1000
9.	254	24.	1000
10.	354	25.	999
11.	454	26.	1000
12.	70	27.	601
13.	170	28.	701
14.	270	29.	711
15.	270	30.	777

209 A & B			Second Half
1.	3	11.	293
2.	13	12.	283
3.	12	13.	280
4.	112	14.	33
5.	10	15.	33
6.	110	16.	68
7.	99	17.	48
8.	95	18.	28
9.	85	19.	128
10.	185	20.	24

209 A & B			Second Half
1.	100	11.	5
2.	200	12.	15
3.	400	13.	25
4.	20	14.	13
5.	21	15.	27
6.	8	16.	16
7.	28	17.	33
8.	28	18.	23
9.	30	19.	49
10.	25	20.	60

Answers

210 A & B			First Half
1.	3	11.	4
2.	6	12.	3
3.	9	13.	5
4.	12	14.	2
5.	15	15.	4
6.	12	16.	3
7.	24	17.	36
8.	36	18.	72
9.	48	19.	2
10.	72	20.	4

211 A & B			First Half
1.	2	16.	20
2.	2	17.	20
3.	4	18.	8
4.	6	19.	10
5.	6	20.	20
6.	8	21.	18
7.	10	22.	16
8.	12	23.	14
9.	12	24.	12
10.	14	25.	8
11.	14	26.	6
12.	16	27.	4
13.	16	28.	0
14.	18	29.	0
15.	18	30.	16

212 A & B			First Half
1.	3	16.	0
2.	6	17.	30
3.	3	18.	27
4.	6	19.	21
5.	9	20.	15
6.	12	21.	12
7.	18	22.	18
8.	15	23.	10
9.	15	24.	9
10.	18	25.	7
11.	21	26.	5
12.	24	27.	3
13.	21	28.	8
14.	27	29.	6
15.	30	30.	1

210 A & B			Second Half
1.	6	11.	1
2.	3	12.	2
3.	12	13.	4
4.	9	14.	1
5.	15	15.	2
6.	12	16.	3
7.	36	17.	36
8.	24	18.	72
9.	48	19.	2
10.	60	20.	4

211 A & B			Second Half
1.	0	16.	20
2.	0	17.	20
3.	2	18.	8
4.	4	19.	10
5.	6	20.	20
6.	8	21.	18
7.	8	22.	16
8.	12	23.	14
9.	12	24.	12
10.	14	25.	8
11.	14	26.	6
12.	16	27.	4
13.	16	28.	0
14.	18	29.	0
15.	18	30.	16

212 A & B			Second Half
1.	0	16.	0
2.	3	17.	30
3.	0	18.	27
4.	3	19.	21
5.	6	20.	15
6.	12	21.	12
7.	21	22.	18
8.	18	23.	10
9.	18	24.	9
10.	15	25.	7
11.	18	26.	5
12.	21	27.	3
13.	18	28.	8
14.	24	29.	6
15.	27	30.	1

Answers

213 A & B			First Half
1.	2	16.	24
2.	4	17.	14
3.	9	18.	12
4.	6	19.	18
5.	3	20.	21
6.	12	21.	27
7.	12	22.	27
8.	18	23.	18
9.	15	24.	20
10.	10	25.	30
11.	12	26.	27
12.	8	27.	21
13.	14	28.	24
14.	21	29.	18
15.	16	30.	12

214 A & B			First Half
1.	1	11.	15
2.	3	12.	15
3.	2	13.	15
4.	2	14.	7
5.	4	15.	17
6.	4	16.	17
7.	2	17.	16
8.	2	18.	18
9.	6	19.	58
10.	6	20.	16

215 A & B			First Half
1.	129	11.	110
2.	147	12.	90
3.	327	13.	92
4.	27	14.	82
5.	54	15.	120
6.	40	16.	120
7.	50	17.	140
8.	90	18.	122
9.	80	19.	146
10.	100	20.	156

213 A & B			Second Half
1.	1	16.	18
2.	2	17.	12
3.	4	18.	14
4.	6	19.	21
5.	12	20.	18
6.	9	21.	24
7.	15	22.	27
8.	21	23.	18
9.	18	24.	20
10.	12	25.	30
11.	14	26.	27
12.	18	27.	21
13.	27	28.	24
14.	24	29.	18
15.	16	30.	12

214 A & B			Second Half
1.	1	11.	15
2.	2	12.	15
3.	2	13.	15
4.	2	14.	7
5.	5	15.	17
6.	5	16.	17
7.	3	17.	16
8.	3	18.	18
9.	5	19.	58
10.	5	20.	16

215 A & B			Second Half
1.	127	11.	110
2.	145	12.	90
3.	325	13.	92
4.	29	14.	82
5.	56	15.	120
6.	70	16.	120
7.	70	17.	140
8.	80	18.	122
9.	90	19.	148
10.	100	20.	158

Answers

216 A & B		First Half	
1.	18	11.	177
2.	23	12.	187
3.	32	13.	196
4.	145	14.	613
5.	144	15.	274
6.	154	16.	621
7.	145	17.	804
8.	165	18.	413
9.	171	19.	417
10.	191	20.	427

217 A & B		First Half	
1.	16	16.	98
2.	19	17.	33
3.	29	18.	43
4.	83	19.	53
5.	59	20.	54
6.	89	21.	64
7.	75	22.	67
8.	87	23.	66
9.	89	24.	60
10.	99	25.	63
11.	98	26.	64
12.	98	27.	66
13.	98	28.	73
14.	98	29.	81
15.	98	30.	94

218 A & B		First Half	
1.	23	11.	185
2.	33	12.	171
3.	42	13.	292
4.	53	14.	337
5.	65	15.	305
6.	64	16.	310
7.	65	17.	306
8.	75	18.	311
9.	76	19.	301
10.	84	20.	311

216 A & B		Second Half	
1.	17	11.	166
2.	22	12.	176
3.	42	13.	196
4.	142	14.	613
5.	145	15.	274
6.	155	16.	621
7.	146	17.	804
8.	166	18.	413
9.	152	19.	417
10.	162	20.	427

217 A & B		Second Half	
1.	15	16.	86
2.	18	17.	31
3.	28	18.	41
4.	85	19.	51
5.	59	20.	52
6.	90	21.	62
7.	78	22.	64
8.	88	23.	63
9.	89	24.	60
10.	99	25.	63
11.	98	26.	64
12.	98	27.	66
13.	98	28.	73
14.	98	29.	81
15.	98	30.	94

218 A & B		Second Half	
1.	24	11.	186
2.	34	12.	152
3.	44	13.	290
4.	54	14.	335
5.	64	15.	305
6.	63	16.	310
7.	65	17.	306
8.	75	18.	311
9.	76	19.	301
10.	84	20.	311

Answers

219 A & B		First Half	
1.	10	11.	125
2.	13	12.	144
3.	103	13.	192
4.	105	14.	135
5.	104	15.	161
6.	106	16.	185
7.	129	17.	193
8.	131	18.	188
9.	144	19.	186
10.	144	20.	198

220 A & B		First Half	
1.	199	11.	208
2.	259	12.	122
3.	261	13.	329
4.	336	14.	735
5.	335	15.	595
6.	355	16.	300
7.	410	17.	957
8.	304	18.	449
9.	406	19.	301
10.	513	20.	596

221 A & B		First Half	
1.	4	11.	61
2.	74	12.	261
3.	574	13.	297
4.	538	14.	4
5.	178	15.	304
6.	36	16.	620
7.	336	17.	550
8.	318	18.	560
9.	138	19.	630
10.	60	20.	640

219 A & B		Second Half	
1.	10	11.	135
2.	12	12.	146
3.	102	13.	192
4.	104	14.	135
5.	103	15.	161
6.	105	16.	185
7.	119	17.	193
8.	130	18.	188
9.	134	19.	186
10.	144	20.	198

220 A & B		Second Half	
1.	198	11.	218
2.	258	12.	135
3.	260	13.	349
4.	334	14.	735
5.	333	15.	595
6.	355	16.	300
7.	410	17.	957
8.	304	18.	449
9.	406	19.	301
10.	513	20.	596

221 A & B		Second Half	
1.	5	11.	71
2.	75	12.	281
3.	565	13.	298
4.	529	14.	5
5.	169	15.	305
6.	34	16.	630
7.	334	17.	450
8.	318	18.	560
9.	138	19.	630
10.	60	20.	640

Answers

222 A & B			First Half
1.	1	16.	11
2.	11	17.	10
3.	21	18.	9
4.	201	19.	20
5.	12	20.	19
6.	120	21.	29
7.	12	22.	42
8.	60	23.	40
9.	44	24.	40
10.	22	25.	39
11.	22	26.	30
12.	21	27.	31
13.	11	28.	7
14.	70	29.	7
15.	60	30.	27

223 A & B			First Half
1.	0	11.	104
2.	1	12.	205
3.	101	13.	206
4.	201	14.	207
5.	202	15.	308
6.	302	16.	409
7.	401	17.	251
8.	602	18.	154
9.	603	19.	618
10.	702	20.	325

224 A & B			First Half
1.	12	16.	27
2.	18	17.	6
3.	27	18.	9
4.	12	19.	12
5.	21	20.	21
6.	24	21.	18
7.	12	22.	24
8.	18	23.	27
9.	27	24.	12
10.	6	25.	15
11.	12	26.	9
12.	15	27.	9
13.	24	28.	15
14.	9	29.	27
15.	18	30.	27

222 A & B			Second Half
1.	1	16.	21
2.	11	17.	20
3.	21	18.	19
4.	201	19.	30
5.	12	20.	29
6.	120	21.	39
7.	22	22.	43
8.	50	23.	40
9.	44	24.	40
10.	22	25.	39
11.	22	26.	30
12.	21	27.	31
13.	11	28.	7
14.	70	29.	7
15.	60	30.	27

223 A & B			Second Half
1.	0	11.	203
2.	1	12.	206
3.	101	13.	207
4.	201	14.	208
5.	102	15.	309
6.	302	16.	408
7.	403	17.	351
8.	502	18.	165
9.	603	19.	617
10.	702	20.	314

224 A & B			Second Half
1.	9	16.	24
2.	15	17.	6
3.	24	18.	9
4.	12	19.	12
5.	18	20.	18
6.	27	21.	18
7.	15	22.	24
8.	18	23.	27
9.	27	24.	12
10.	6	25.	15
11.	12	26.	9
12.	15	27.	9
13.	24	28.	15
14.	9	29.	27
15.	18	30.	27

Answers

225 A & B			First Half
1.	4	11.	36
2.	8	12.	40
3.	12	13.	0
4.	4	14.	36
5.	8	15.	32
6.	12	16.	28
7.	20	17.	24
8.	24	18.	40
9.	28	19.	36
10.	32	20.	16

226 A & B			First Half
1.	1	16.	3
2.	1	17.	7
3.	1	18.	5
4.	4	19.	9
5.	3	20.	6
6.	2	21.	10
7.	2	22.	10
8.	4	23.	7
9.	3	24.	8
10.	4	25.	4
11.	5	26.	6
12.	7	27.	2
13.	10	28.	3
14.	8	29.	9
15.	9	30.	9

227 A & B			First Half
1.	2	16.	8
2.	4	17.	16
3.	4	18.	10
4.	12	19.	20
5.	12	20.	12
6.	16	21.	24
7.	16	22.	14
8.	2	23.	28
9.	4	24.	16
10.	12	25.	32
11.	12	26.	18
12.	16	27.	36
13.	16	28.	20
14.	6	29.	40
15.	12	30.	200

225 A & B			Second Half
1.	0	11.	28
2.	4	12.	40
3.	8	13.	0
4.	8	14.	12
5.	12	15.	32
6.	16	16.	28
7.	24	17.	24
8.	28	18.	40
9.	32	19.	36
10.	36	20.	16

226 A & B			Second Half
1.	1	16.	3
2.	1	17.	7
3.	1	18.	5
4.	4	19.	9
5.	3	20.	6
6.	2	21.	10
7.	2	22.	10
8.	4	23.	7
9.	3	24.	8
10.	4	25.	4
11.	5	26.	6
12.	7	27.	2
13.	10	28.	3
14.	8	29.	9
15.	9	30.	9

227 A & B			Second Half
1.	0	16.	8
2.	2	17.	16
3.	0	18.	10
4.	12	19.	20
5.	16	20.	12
6.	16	21.	24
7.	16	22.	14
8.	4	23.	28
9.	8	24.	16
10.	12	25.	32
11.	9	26.	18
12.	14	27.	36
13.	28	28.	20
14.	12	29.	40
15.	12	30.	200

Math Sprints 2

Answers

228 A & B			First Half	
1.	0¢	16.	85¢	
2.	35¢	17.	65¢	
3.	80¢	18.	55¢	
4.	75¢	19.	35¢	
5.	50¢	20.	15¢	
6.	25¢	21.	5¢	
7.	95¢	22.	25¢	
8.	70¢	23.	95¢	
9.	40¢	24.	5¢	
10.	60¢	25.	90¢	
11.	30¢	26.	10¢	
12.	10¢	27.	0¢	
13.	20¢	28.	85¢	
14.	15¢	29.	65¢	
15.	25¢	30.	55¢	

229 A & B			First Half	
1.	$4.75	11.	$5.50	
2.	$4.50	12.	$5.95	
3.	$10.25	13.	$5.15	
4.	$7.85	14.	45¢	
5.	$1.00	15.	70¢	
6.	$2.00	16.	$3.70	
7.	$1.00	17.	$5.25	
8.	$6.00	18.	$1.50	
9.	$1.00	19.	$4.55	
10.	$5.00	20.	$7.95	

228 A & B			Second Half	
1.	0¢	16.	85¢	
2.	90¢	17.	65¢	
3.	80¢	18.	55¢	
4.	75¢	19.	35¢	
5.	50¢	20.	15¢	
6.	25¢	21.	5¢	
7.	95¢	22.	25¢	
8.	70¢	23.	95¢	
9.	40¢	24.	5¢	
10.	60¢	25.	90¢	
11.	30¢	26.	10¢	
12.	10¢	27.	0¢	
13.	20¢	28.	85¢	
14.	15¢	29.	65¢	
15.	25¢	30.	55¢	

229 A & B			Second Half	
1.	$2.75	11.	$2.50	
2.	$3.50	12.	$4.95	
3.	$6.25	13.	$2.15	
4.	$7.85	14.	$2.45	
5.	$1.00	15.	70¢	
6.	$2.00	16.	$4.70	
7.	$1.00	17.	$4.25	
8.	$5.75	18.	$1.50	
9.	$1.00	19.	$4.55	
10.	$4.00	20.	$7.95	